T0220489

Individual and Collective Memory Consolidation

Individual and Collective Memory Consolidation

Analogous Processes on Different Levels

Thomas J. Anastasio, Kristen Ann Ehrenberger, Patrick Watson,
and Wenyi Zhang

The MIT Press
Cambridge, Massachusetts
London, England

© 2012 Massachusetts Institute of Technology

All rights reserved. No part of this book may be reproduced in any form by any electronic or mechanical means (including photocopying, recording, or information storage and retrieval) without permission in writing from the publisher.

This book was set in Stone Sans Stone Serif by Toppan Best-set Premedia Limited.

Library of Congress Cataloging-in-Publication Data

Individual and collective memory consolidation : analogous processes on different levels / Thomas J. Anastasio ... [et al.].
  p. cm.
Includes bibliographical references and index.
ISBN 978-0-262-01704-6 (hbk. : alk. paper), 978-0-262-54400-9 (pb.)
1. Memory. 2. Collective memory. 3. Identity (Psychology) 4. Group identity.
I. Anastasio, Thomas J.
BF371.I53   2012
153.1'2—dc23
2011028995

Contents

Contents

Preface

Interdisciplinary contexts foster creative work because they inspire research-ers to carry concepts analogically across domains of knowledge (see, e.g., Gardner, 1982, 1993; Boden, 1990; Weisberg, 1993; Feldman et al., 1994; Csikszentmihalyi, 1996; Dunbar, 1997; Otis, 2001, 2002). Pioneering inter-disciplinarian Margaret Boden (1990) suggests that creative ideas can be generated by representing a problem as an analogy between two different systems or phenomena. Our goal was to generate creative ideas about memory formation by making an analogy between the processes of memory consolidation as they occur on individual and collective levels. Our work began in an interdisciplinary context.

At the University of Illinois, the Illinois Program for Research in the Humanities and the Center for Advanced Study examined memory in academic year 2004–2005 with "The Memory Project: An Interdisciplinary Study of Memory and the Construction of Identity and Culture," organized by Bill Brewer and Lillian Hoddeson. Under these auspices, a weekly seminar in fall 2004 brought faculty and graduate students together from disciplines as diverse as English and mathematics, computer science and education, music and psychology, to learn from and inspire each other. The outcome of that collective endeavor was a broadening of each partici-pant's conceptualization of the nature of memory and the realization that no one discipline could possibly claim to address the subject of memory in its entirety.

To pursue the relationship between individual and collective memory in greater depth, Thomas J. Anastasio and Lillian Hoddeson conceived the Memory Analogies Group at the University of Illinois. With funding from the Beckman Institute, this initiative has enabled a small interdisciplinary group to pursue analogies between what neuroscientists currently under-stand about memory formation in individual brains and what humanists understand about the making of collective memories. In addition to faculty members Anastasio (neuroscience) and Hoddeson (history), the team consisted of graduate students Kristen Ann Ehrenberger (history), Patrick Watson (psychology), and Wenyi Zhang (anthropology).

In our research, we have tried to heed French philosopher Roland Barthes' caution that our "Interdisciplinary studies . . . do not merely confront already constituted disciplines" such that, rather than merely comment on the views from various fields, we have tried to create "a new object, which belongs to no one" (Barthes, 1984/1986, p. 72). Many scholars have pleaded for dialogue among the various disciplines that deal with individual or collective memory, but they have—in practice—treated the two as sometimes connected but otherwise effectively incommensurable. In contrast to previous work, we face this challenge full on and argue that memory formation in individuals is analogous to memory formation in collectives and that the processes that underlie memory formation perform essentially the same functions on various levels from a single brain to a group, community, or nation. Our "new object," and the subject of this monograph, is a unique and thoroughly interdisciplinary framework for understanding what is known in neuroscientific circles as "consolidation," the process by which new memories form. In brief, we contend that the consolidation of memory is the (often biased) process by which individuals and collectives hold, select, relate, generalize, and stabilize new and fleeting memory items into more or less permanent representations of knowledge. The ways in which they do so, we will argue, are different but analogous.

Our interdisciplinary analysis will provide new perspectives on both sides of C.P. Snow's (in)famous two cultures divide (Snow, 1959). For example, memory scientists who have tried to ignore the personal idiosyncrasies of their individual subjects will benefit from a fact of which humanists have long been aware: that the goals, desires, and established memories of the remembering entity, whether an individual or a group, can influence (bias) the consolidation of new memories. Humanists, meanwhile, have been accused of favoring description over explanation. Studies abound that address socially and culturally constructed collective memories, but in the humanities there is little appreciation of consolidation as a process in itself. Humanists will benefit from the very concept of consolidation, which explains how memories form (consolidate) so that individuals as well as social groups can remember them later.

Having combined theorization with directed literature surveys, we hope that our interdisciplinary conceptualization of the consolidation of memory in individuals and groups will be useful to researchers working in an academic environment that is admittedly more conducive to parallel than to integrated research. Rather than present four separate views of individual or collective memory consolidation, we have made every effort to synthesize our various contributions into one seamless but multidisciplinary whole. This monograph is as much a single work as one faculty member and three very independently minded graduate students could possibly make it. It is our ambition that humanists and scientists alike will find value in this monograph and can use it to enrich the original work they are already doing by making analogies to research on memory in an interdisciplinary fashion.

Acknowledgments

The authors wish to acknowledge the Beckman Institute for the grant that supported this project from academic year 2006 through academic year 2011. We especially wish to thank Lillian Hoddeson for her help in setting this project in motion and for her encouragement along the way. Many thanks also go to our scholarly consultants for generously sharing their time and knowledge: Bill Brewer and Neal Cohen in cognitive psychology, Kai-Wing Chow and Peter Fritzsche in history, Janet Keller and F. K. Lehman in anthropology, and Michael Rothberg in English. Many thanks also to Walton Kelly in geology for his comments on the manuscript. We are very appreciative of the positive responses we received when parts of the monograph were presented at the 2007 Thinking Affect: Memory, Language, and Cognition conference organized by Jennifer Lieberman and Elizabeth Hoiem for The Unit for Criticism and Interpretive Theory at the University of Illinois at Urbana-Champaign and when the entire monograph was presented in summary form in 2008 at the Microsymposium on Individual and Collective Memory, organized by Anastasio and held at the Beckman Institute. We also wish to express our thanks and appreciation to our respective families for their love and support during the adventure of this interdisciplinary collaboration.

1 Introduction

It is curious, that the analogy between the individual and social organization has most frequently been used by those who, having already decided what individual characteristics are important, wish to transfer these to the group. Obviously it can equally well be turned round, for whatever the argument is worth we could use it just as well to urge that the individual is nothing more than a special sort of group, as to maintain that the group possesses all the characteristics of the individual.
—Frederic Bartlett, *Remembering* (1932, p. 309)

British psychologist Frederic Bartlett (1886–1969) is best known today for postulating the existence of "schema," generalized frameworks that organize knowledge and aid recall. Schema are now a widely accepted characteristic of human memory, but when *Remembering* was published in 1932 the book received little attention, even though Bartlett was a highly respected Cambridge don. Bartlett also had a background in anthropology, and he wanted to study the influence of cultural groups on individuals' memories. His fellow psychologists, however, were heavily invested in the study of isolated individual behaviors, so Bartlett's ideas were not translated into experimental practice or included in conceptualizations of memory formation.

During the "cognitive revolution" of the 1950s and 1960s, scientists finally got around to reading *Remembering*. They embraced the idea of schemata as components of individual memory but they concentrated only on the first half of the book. The second half of *Remembering*, which Bartlett had entitled "Remembering as a Study in Social Psychology," was more or less ignored by cognitivists and by social psychologists, whose disciplines developed independently from sociology in the mid-twentieth century. Apparently, no one applied Bartlett's framework in the study of collective memory formation. In this book, we retake the path Bartlett cut more than seven decades ago, and we intend to clear it and widen it by arguing that individual and collective memory formation are examples of analogous processes that occur on different levels. Further, we intend to tread this path in both directions and will argue that insights gained on one level can be applied on the other.

Across the decades of research on memory in (and of) individuals and groups that has been performed since 1932, Bartlett's first observation in the quotation above—that most often it is individual characteristics that are "transferred" to groups—has remained spot on. "Collective memory," for instance, is frequently described as just such a property borrowed from individual neuroscience (e.g., Novick, 1999; Kansteiner, 2002, 2007), where memory is broadly defined as the ability to form long-lasting knowledge constructs that, once formed, can be efficiently accessed. For our part we have, in fact, attributed to social groups the same process of memory formation that has been described for individual memory. However, we have also "turned" the direction of attribution "round," and borrowed from the social sciences and humanities an emphasis on the influence of what we call "the entity." For our purposes an entity is an individual, family, community, nation: any person or group that possesses the properties of memory. Even social entities, which have "collective memory," must be able to form it, and we argue that individual and collective memories are formed through analogous processes. For individual memory, that process has been termed *consolidation*.

Consolidation refers to the conversion of more immediate and fleeting bits of information into a stable and accessible representation of facts and events, including a representation of the world and the entity's place in it. For individuals and collectives alike, memory enables more effective interaction with the world and serves as part of their basis for decision and action. For our purposes, three aspects of memory should be distinguished: consolidation, remembering, and structures. Memory consolidation (i.e., formation) produces memory structures that can then be used for remembering. Both consolidating and remembering are processes, but in this book we concern ourselves primarily with memory consolidation. Although memory structures are different on individual and collective levels, we maintain that, on an abstract level, the processes of individual and collective memory consolidation are analogous.

Stated most simply, memory consolidation can be thought of as the process that transforms *short-term* memories into *long-term* memories. Short-term, or unconsolidated, memories are *labile*, meaning that they are disruptable. Long-term, or consolidated, memories are *stable*—changeable but persistent. Short-term memories, long-term memories, and the processes of consolidation are all associated with various structures on individual and collective levels. On the individual level, the structures that enable memory storage and processing are, of course, neurons and their interconnections, as organized into various brain regions. On collective levels, these structures include museums, monuments, books, newspapers, and people, as organized into various groups such as congresses, communities, and courts. Importantly, it is not a memory structure in itself but the way it fits into the organization of memory that determines whether it is short term or long term. Similarly, and assuming he is still with us, your father's mustache exists as part of the (long-term) organization of his face, despite the fact that the actual (short-term) hairs turn over every time he trims it. Whatever the structures associated with them, the short-term inputs to the consolidation process can include isolated facts and sequences of occurrences, and the long-term outputs can include schemata, stories, narratives, paradigms, and frameworks, into which specific facts and episodes can be inserted. As individual consolidation is a demonstrable phenomenon, and as both individuals and collectives exhibit memory, we take the radical step of postulating a single process of memory consolidation that manifests in analogous ways on the individual as well as on collective levels.

Languages of Memory

The term "memory" has had so many different uses in such disparate fields—from electronic computing to human learning, materials science to collective remembering—that its meaning today is very complicated. To begin with, the language used to describe what the brain does is older than the study of the brain itself. Already in ancient times, authors wrote about Mnemosyne ("Memory"), the mother of the Muses, and made observations about how memory might work (Yates, 1966; Herrmann & Chaffin, 1988). The most famous ancient empiricist of memory is Simonides of Ceos (ca. 556–468 BCE), whom Cicero (106–43 BCE) credits with the *ars memoriae*, or "art of memory," after the mnemonic he proposed. Memory was putatively localized during the Middle Ages, when natural philosophers believed that the brain was simply padding or a heat sink for the ventricles that actually contained thoughts. The rear-most ventricle supposedly contained *memoratius*, "reflection upon the past." Because of the proliferation of research since then, the term "memory" has gained new associations and no longer correlates cleanly with a single phenomenon. In neuroscience, "memory" has become an umbrella term under which congregate myriad phenomena. For the purposes of our consolidation project, we confine ourselves to "declarative memory," the type of memory that concerns the storage and retrieval of knowledge that can be expressed in words or some other demonstrable fashion (see chapter 2).

Conversely, although observers of society have been writing about collective thought processes for more than 200 years, the term "memory" is relatively new to the academic study of history (K.L. Klein, 2000), sociology (Olick & Robbins, 1998), anthropology (Birth, 2006a, 2006b; G. White, 2006), and literature (Nalbantian, 2003). Memory in the humanities nevertheless brings with it a lot of baggage. Before "memory" became a buzzword in the 1980s, historians discussed "myth," "tradition," and "legend" (Yerushalmi, 1982; Gedi & Elam, 1996), and because it has become so popular, some wonder whether "memory" has not become such a catch-all that it has lost its usefulness (e.g., K.L. Klein, 2000). Some historians have already begun to call what they study "collective remembering" (Wertsch, 2002) or "collective remembrance" (Winter & Sivan, 1999), because these name an active process. But remembering is not the only process of (collective) memory—there must be a process of (collective) memory formation as well, and some historians are beginning to describe this process (Schwartz, 2000, 2008; Rosenberg, 2003). We will contribute to this discussion with our analogical analysis of "collective consolidation."

The term "memory" as a noun implies a "thing," like an integrated circuit mounted on a card that permits random access by a computer to stored information. With specialized knowledge and equipment, random-access computer memory can be directly probed. Likewise, using electrodes, amplifiers, and other methods, neuroscientists attempt to study "memory" in various forms on the neural level. Psychological memory cannot be probed directly, but it can be assessed indirectly by studying memory storage and retrieval. That approach was taken by Bartlett and his contemporaries and by legions of psychologists after them who have described memory on the individual level. Paradoxically, on collective levels, where "memory" seems the most nebulous, the actual structures of memory are the most accessible—a museum, for example, is a collective memory structure you can walk through (Kavanagh, 1996). Although methods exist for systematically exploring social constructs such as culture (e.g., Quinn, 2005), reports on individual memory are comparatively more quantitative, whereas those on collective memory are more qualitative. Still, the qualitativeness of humanistic work is appropriate to the material available for humanists to study, and we accord the humanistic work a value equal to that of the scientific work in our exploration of individual and collective memory consolidation. Our approach is to take advantage of the knowledge accumulated by both scientists and humanists. We will show that the findings and interpretations of neuroscientists, psychologists, anthropologists, historians, and others on individual and collective memory, derived from their analyses of remembering and of memory structures, are consistent with our model that describes memory consolidation on individual and collective levels.

Those who study consolidation proper use two definitions of the term, which can be paraphrased from the *Oxford English Dictionary* definition for the verb form "to consolidate": the first is to make strong or solid, the second to combine separate items into a single whole. Neuropsychologists can be understood as having emphasized the first meaning—to solidify or strengthen—to explain why certain amnesics cannot remember people encountered or events experienced shortly before the trauma that caused their amnesia. Damage to a brain structure known as the hippocampus produces two types of amnesia: retrograde and anterograde (Manns et al., 2003). Anterograde amnesia is the inability to form new long-term memories, and retrograde amnesia is the loss of previously held short-term memories. As more than a century of investigations into retrograde amnesia demonstrate, long-term memories take time to form, and if some trauma disrupts this process, then the "unsolidified," or unconsolidated, memories will be lost, but "solidified," or already consolidated, memories will be spared. The relative sparing of already consolidated (or long-term) memories is a characteristic feature of retrograde amnesia in individuals. Part III of this book provides evidence that this form of retrograde amnesia also occurs on collective levels as the result of specific forms of societal trauma.

Although the fact of retrograde amnesia has provided the bulk of the evidence for the consolidation process as one of strengthening memory "traces," it does not explain how the process itself works. In other words, retrograde amnesia, with its relative loss of more recent (and not yet consolidated) memory, provides negative proof of a process necessary to create memories. This is where the second definition of consolidation—to combine into a whole—applies, and again, not only in individuals but also in collectives. Consolidation in the sense of combining involves both labile (short-term) and stable (long-term) memory, with parts of each being selected and related to create new, or to modify existing, long-term memory structures. Rather than simply gathering memory items together, the consolidation process so-defined produces cognitive structures that organize memory items and facilitate both memory storage and its retrieval. Because brains and societies have access to both consolidated (long-term) and unconsolidated (short-term) knowledge, individual and collective memories are always up to date and capable of responding to the present moment. Memory formation is thus not an end in itself but an ongoing process that is part of the whole of "memory."

Stable memory is long term but not rigid or unchangeable, just as your father's mustache can be a long-term part of his look but still change in style or go gray. The consolidation process dynamically structures the knowledge that individuals and collectives extract from their environments and makes it consistent with their ongoing experience. Conversely, what has been previously learned can affect the formation of new memories, because the knowledge held in long-term stores can influence the selection and relation of short-term memory items, both with each other and with existing, long-term memories. This recursive characteristic of memory consolidation will resurface from time to time throughout our text.

A Memory Analogy

We use a definition of *analogy* that is different than the one some other scholars use. Comparative literary scholar Laura Otis (1999, 2002), for instance, uses "metaphor" the way we use "analogy." She describes how, in the 19th century, neuroscientists and communications technologists shared knowledge with each other in the form of metaphors [sic]: "Just as [Hermann] Helmholtz saw nerves as telegraphs, [Samuel] Morse and other designers of the telegraph saw their wires as nerves" (Otis, 2002, p. 119). In describing neural and telegraph systems in the same language, they talked about "wires" and "nerves" not for oratorical flair, but because that was how they understood their disciplines—one by way of the other. Both fields benefited from this exchange. But, as Otis points out, "The inability of nerves to resume activity when spliced . . . indicated a key difference between organic and technological communications systems. They were analogous, but not identical" (p. 114). Nerves are not wires after all, so we would explain the situation this way: Although these scientists and technologists at first believed their systems to be analogical (i.e., interchangeable), it turned out that they are in fact metaphorical (only superficially similar).

Good metaphors allow us to look at something in a new way, and they can be marvelously descriptive. We will also use metaphors extensively throughout this monograph to help us describe various aspects of memory consolidation. But metaphors are not explanatory, because the two terms are, in the end, incommensurable. A metaphorical relationship is unidirectional: one could say, "a knife of lightning cut the sky," but one would not say, "I used a lightning knife to cut the turkey." In other words, we are grouping metaphors with similes instead of with analogies, according to their underlying meanings. So the lightning is like a knife, but a knife is not like lightning (although perhaps a person could cut *as fast as* lightning).

By contrast, analogous relationships are bidirectional, and it is this interchangeability that allows for their explanatory power. French philosopher Georges Canguilhem (1904–1995) used metaphor and analogy the way we do to characterize how nineteenth- and early-twentieth-century biomedical researchers used concepts borrowed from other disciplines, such as pathological anatomy, to describe phenomena for which they could not yet explain the underlying mechanism. He writes, "At the outset, the concept of hereditary biochemical error rested on the ingenuity of a metaphor" borrowed from pathological anatomy; "today it is based on the solidity of an analogy" that recognizes what exactly has gone wrong to cause a metabolic disorder (Canguilhem, 1966/1978, p. 172). In other words, biochemistry has a physical basis just like anatomy does, such that enzymes can, by analogy with gross morphology, be the seat of pathology. Metaphors are colorful but unidirectional and descriptive, whereas analogies based on knowledge of the underlying structures and especially functions allow one to establish a more precise, bidirectional, and explanatory relationship.

The weak version of our hypothesis would be the metaphor that collective memory consolidation *is like* individual memory consolidation. The strong version is an analogy: that the underlying concept of "memory consolidation" is the same on both—or more precisely, on all—levels, such that memories are created via the same process (consolidation) using the means appropriate for each level (the individual brain and various social structures). We argue the latter, that individual and collective memory consolidation processes are analogous, because they display characteristics that are conceptually interchangeable. Specifically, we will describe the creation of memories in both individuals and collectives with the same procedural steps: buffered storage, selection and relation, generalization and specialization (see next section). Thus, it is possible to learn about individual memory consolidation by studying collective memory consolidation, and vice versa.

The same principle of analogy is at work, for instance, with the concept of a negative feedback loop. If a negative-feedback system senses a deviation of its controlled variable from a set point, it drives an actuator to counteract that change and brings its controlled variable back in line with the set point. That is the case with the homeostatic regulation of body temperature by the hypothalamus and with the thermostatic regulation of the temperature in a house. No one would argue that the hypothalamus and a thermostat are physically similar: the hypothalamus functions neuroendocrinologically, the thermostat electronically. Yet it is evident that both systems regulate the same thing—temperature—and they regulate it in the same way, by returning the perturbed system to a set point. Likewise with memory: Societies and individuals both form memories, and while the embodiments are different and appropriate to the various entities, the underlying principle—in our case, consolidation—is the same. So whereas individuals consolidate memories via reorganization of neural systems, collectives consolidate memories in the creation and propagation over space and time of stories, facts, and myths about the world, themselves, and *wie es eigentlich gewesen* ("the way things really were"). The structures may be different, but the process is the same.

We will frequently refer to "various levels of memory" as a shorthand way of acknowledging that memory is studied at many magnifications, or "levels" of abstraction. *Individual memory* encompasses synaptic, neuronal, brain, and psychological levels. *Collective memory* refers to supra-individual levels: couple, family, community, nation, religion, and so forth. The collective categories may be hierarchical (city, region, state) or horizontal. Horizontal relationships occur at the same level of abstraction—namely groups of individuals—and frequently overlap, as is the case with most social memberships. For instance, it is possible to belong to a profession, a nation, and a religion simultaneously; none of these groups is reducible to another. Maurice Halbwachs (1877–1945), who is widely considered to be the father of collective memory studies, put it this way: "the great majority of these groups, even though not currently divided, nevertheless represent, as Leibniz said, a kind of social material indefinitely divisible in the most diverse directions" (Halbwachs, 1950/1980, p. 85). Without going into the combinatorics, we acknowledge that, in a group of any size, the number of potential subgroups is extremely large. In a real society there are many social subgroups with overlapping membership, as well as a hierarchy of subgroups. Each subgroup can influence the others, and all can have their own collective memories. Because we focus on declarative memory, which is a whole-brain phenomenon, we are concerned mainly with the single, whole-brain individual level but we consider multiple collective levels. Our model of memory consolidation applies on the (whole-brain) individual level and on all collective levels.

Neither is the distinction we make between the individual and collective levels meant to imply that they are functionally separate: we acknowledge an essential interaction. In his seminal work *On Collective Memory* (Halbwachs & Coser, 1925/1992), Maurice Halbwachs argues that individual memory is socially mediated and that "social frameworks" determine what an individual remembers and how she remembers it. It is possible to conceive of individual memory that occurs independently of society. In neurobiology, memory is the result of learning, and learning is any experience-dependent change in behavior. For example, a sea-slug can learn to reduce the amount by which it withdraws its gill in response to repeated but innocuous squirts of water (Pinsker et al., 1970). It is hard to argue that this behaviorally adaptive "memory" of previous stimuli is socially mediated. In contrast, autobiographical writing, a seeming act of individual memory formation, is strongly socially mediated and could be considered as an act of collective memory formation (e.g., Berntsen & Bohn, 2009).

We will take this interaction between levels into account as we draw a distinction between individual and collective memory consolidation, which we must do so that we can compare them and argue that they are analogous processes. The physicist Per Bak (1996) did the same when he showed that the movement of grains of sand on a beach is analogous to the movement of the tectonic plates that make up the earth's crust. Tectonic plates are composed of particles of rock that are like sand, and it is certain that grains of sand on a beach will move during an earthquake, but the interlevel relationships between sand and plates do not invalidate the analysis showing that the processes occurring on their respective levels are analogous. We certainly do not deny interactions between individual and collective levels of memory formation, but we do make a distinction between these levels to show that they are analogous.

By making analogies between individual and collective memory—which is *roughly* synonymous with drawing parallels between memory as described in the sciences and in the humanities—we were able to construct an interdisciplinary, hybrid theory of the process of the stabilization of memories that could not have been conceived by studying only one level of memory in a single field of research. In this way, we transgressed C.P. Snow's "two cultures divide" (Snow, 1959), but in so doing we took a risk. Not only are the "two cultures" not supposed to meet, but also many investigators are invested in separating individual from collective memory. Most scientists demur at the notion of collective memory, as understanding individual memory poses sufficient challenge. Many humanists agree: Everything is socially mediated anyway, so one might as well look only at individuals and not at some phantom "group" that erases individual agency. By contrast, we are invested in interweaving findings on individual *and* collective memory. What we are trying to suggest is that—with whatever disciplinary tools you use to study individual or collective memory—you consider that there is a single process of consolidation with analogous manifestations on any level, whether that of a brain, a family, or a society. We found the risk worth taking, and we hope you will find that the data and ideas that have accumulated on individual and collective memory fit nicely into our framework.

Finally, there is a certain pragmatism to our interdisciplinary comparison. Because collective experiences occur on the level of interpersonal interaction, they can be observed more easily than changes occurring within individual brains. This advantage is balanced by the drawback that collective memory can be extraordinarily complicated. If memory in a single individual were not difficult enough to define and measure, collective consolidation and remembering require sometimes very large numbers of individuals whose beliefs and actions often break down into a multiplicity of collective memories, which makes definition and measurement all the more difficult. Clearly, the study of memory presents great challenges to researchers on any level. The strategy of comparing individual and collective memory allows us to exploit the advantages of each and, hopefully, to provide greater insight into both.

The Form of the Content

The rest of this book consists of three parts. Part I introduces the reader in separate introductory essays to individual and collective memory and describes our model of memory consolidation. Chapter 2 provides a literature review of scientific work on consolidation in individual memory over the past 100-plus years. It is intended both to locate our study for scientists and to bring humanists up to speed on the process of consolidation as it is currently understood in individual minds and brains. Until now there has been no concept of collective consolidation in the humanities literature, so chapter 3 defines (and defends) what we mean by "collective memory." It is largely directed toward humanists who discredit the idea of collective memory, but it is also written in such a way as to introduce scientists to the long-standing concept of collective memory. Chapter 4 may stand alone as a statement of our thesis: We believe individual and collective memory consolidations are analogous manifestations of the same process on different levels. This chapter lays out our "three-in-one," interdisciplinary theory of consolidation and its constituent elements: buffered storage, selection and relation, generalization and specialization (the "three"), and "the entity" (the "one"), which draws from each of the other elements in remembering and acts on each of the other elements in influencing memory formation.

In part II, we explain in depth each element described in chapter 4 using illustrations from existing individual and collective memory literature. We survey both primary and secondary literature because our aim is to show that not only original findings but also interpretations of findings are consistent with our model of memory consolidation. Although each chapter begins with individual examples before moving on to collective examples, we do not mean to suggest that all of our ideas come from the scientific study of consolidation. Rather, we decided to progress from lower levels of "magnification" (i.e., individuals) to higher levels (i.e., collectives), but this decision was arbitrary.

In chapters 5 and 6, we acknowledge that the initial stages of consolidation are a bit of a juggling act. Chapter 5 points out that consolidation requires temporary "storage and handling" of—and attention to—both short-term (unconsolidated) and long-term (already consolidated) items. From this easily manipulated state, items in the "buffer" must be selected and related to each other in preparation for the rehearsal process that commits them to memory as stable (long-term) representations. Chapter 6 explores how selection and relation are mediated through the hippocampus, a part of the cerebrum that plays a central role in memory consolidation. The set of interrelationships produced through the operation of the hippocampus embodies many meanings, or polysemy, a concept familiar to humanistic scholars and recently broached by memory scientists. Whereas some memory items are indeed stored verbatim as specific facts or episodes, most are combined into general categories, and the relationships between individual items and categories are also stored. In chapter 7, schema and textbooks serve as examples of generalized forms of individual and collective memory, respectively. Generalization and its attendant loss of detail exist in tension with specialization and its retention of detail to produce a system that, when fine-tuned over a long period of time, extracts and stores the knowledge contained in a set of relations and makes it available in a quickly and easily accessible form. Thus, chapters 5, 6, and 7 describe the three core elements of the memory consolidation process.

We have labeled the fourth and final factor in the consolidation process "the entity." "Entity" here serves as a catch-all term, a blank to be filled in with the object of the analysis: in psychology, an individual research subject; in history or anthropology, a particular collective. The environment is the context for an entity, and an entity provides the context for the consolidation process. The entity simultaneously draws from the other three memory elements and influences them, making it a meta-element that works with but is essentially different from the other elements. Chapter 8 explains how an entity, partly by reference to existing knowledge but mostly through its intrinsic goals and desires, can affect the other three elements and thereby influence the consolidation process. Chapter 8 also provides an overview of our model of memory consolidation and, together with its explication of what we mean by "entity effects," complements chapter 4.

In part III, we put our model to the test by deriving and evaluating a prediction. Our model predicts that retrograde amnesia, which is well described on the individual level, should also occur on collective levels. The retrograde amnesia that results from traumatic injury of the hippocampus in individuals is characterized by loss of recent but sparing of remote memories (Squire & Alvarez, 1995). We hypothesize the existence of a "social hippocampus" composed of opinion leaders in various fields who establish the relationships between events that provide the substrate for collective memory formation. We then study the Chinese Cultural Revolution (1966–1976) as an instance of trauma to a social hippocampus. We present evidence for the relative loss of recent collective memory in mainland Chinese populations that experienced the Cultural Revolution compared with Chinese populations off the mainland that did not, and for retention of remote memory in both groups. The correspondence in memory loss patterns between the mainland Chinese, who experienced the Cultural Revolution, and individuals with hippocampal damage strengthens our argument that collective retrograde amnesia occurs as a result of damage to a "social hippocampus."

Chapter 9 defines our hypothesis concerning collective retrograde amnesia and sets the stage for our case study. Our goal is to compare the collective memories of mainland Chinese with those of groups within the Chinese diaspora that were not affected by the Cultural Revolution. In the chapter, we briefly sketch the relevant history of the "Thinking Generation" in mainland China, the Nationalist Chinese in Taiwan, and the refugee Chinese in Northern Thailand. In chapter 10, we explore their collective memories for traditional aspects of Chinese culture, focusing on religion. We find retention of memory for long-established components of religions in all three populations but, among the mainland Chinese who experienced the Cultural Revolution, we find relative loss of memory for religious meaning and for other details that require constant rehearsal to link current experience with the cultural past. In chapter 11, we consider collective memory for more recent aspects of Chinese culture. Among the population that experienced the Cultural Revolution, we find relative loss of memory for recent (early 20th century) literary developments and authors. We also note the rapid recovery of memory on mainland China for this literature, as the mainland has become more open and as intercourse with non-mainland Chinese has increased since the 1980s and 1990s.

We conclude our argument in chapter 12, where we indicate some of the implications of our work for conceptualizations of, and further research on, individual and collective memory. We emphasize the bidirectional nature of our analogy and point out specific ways in which observations on one level can provide new perspectives on another level. For example, our analogy offers memory scholars a new way to view collective memory formation (as multistage consolidation) and offers scientists another angle from which to approach the study of memory distortion (due to the influence of the remembering entity). Although we stress the strengths of our analogy, we also admit its shortcomings, and we end by happily noting that unlike the biological hippocampus, a damaged social hippocampus can be regenerated, and recent memories lost through disruption of collective consolidation can be restored.

I Types of Memory

1 Types of Memory

2 Individual Memory and Forgetting

This chapter provides a brief history of the concept of individual memory consolidation and a survey of some of the psychological and neuroscientific research that has grown out of it. Over the past 100-plus years, the conception of consolidation has slowly changed from the verbatim storage of discrete packets of information ("memory items") into a recursive process involving the formation and continued reformation of more generalized mental representations. Along the way, "consolidation" has accreted entire vocabularies, much in the way that "memory" has become an umbrella term for many different cognitive functions. Our model contributes to the shift toward an understanding of memory consolidation as the construction and continued reorganization of mental representations. In this chapter, we trace the genealogy of consolidation and assess its continuing significance in the science of individual memory.

The First (Psychological) Description of Consolidation

The history of memory consolidation begins in earnest with German psychologist Georg Elias Müller and his student Alfons Pilzecker in 1900 (Müller & Pilzecker, 1900), but understanding their work requires some historical context. Müller and Pilzecker were using an experimental paradigm based on the foundational work of another German psychologist, Hermann Ebbinghaus (1850–1909). Ebbinghaus described this paradigm in his 1885 book *Über das Gedächtnis: Untersuchungen zur experimentellen Psychologie* (*On Memory: Investigations in Experimental Psychology*), usually translated simply as *Memory*. He had the insight and audacity to apply quantitative measurement techniques to a cognitive function, and memory scientists of the empiricist movement soon widely adopted his experimental procedures. These *list-learning techniques* consisted of subjects memorizing lists of nonsense syllables (e.g., zik, bij, yat). After a carefully prescribed delay, the subjects would attempt to recall the list. These techniques quantified memory by measuring the number of rehearsals required for perfect recall of the entire list of syllables or, more commonly, for recall of some portion of the list.

The nonsense syllables used in Ebbinghaus's list-learning experiments might seem to be the atoms of memory, but it is not known whether there is any irreducible unit of memory, or whether memory can be broken down into discrete packets (like yat). It is unlikely that memory is a continuous thing, like a fluid. As we will see, it seems that memory cannot be fully understood without treating its contents as a collection of specific "items," even though the size and structure of those items are difficult at the present time to specify. We will use various metaphors to help us describe memory, sometimes as a set of discrete items and other times as a continuous substance. This is not meant to imply some sort of wave/particle duality for memory but only to acknowledge that the nature of "memory" still defies complete comprehension, and we must use available concepts to describe whatever is known about it. We will also frequently use the term "item" to refer to a member of the set of memory contents, but we will leave unspecified the size and structure of any item.

The nonsense syllable served as the experimental item of choice for early list-learning experiments. Müller and Pilzecker were conducting such experiments when they noticed a phenomenon that would permanently alter the theoretical understanding of memory. If two lists of syllables were given back to back, performance on the first list was poorer than if it was presented by itself (none of the syllables in the first list were repeated in the second). The counterpart to this "retroactive interference" is "proactive interference," by which performance on the second list is also poorer than if it was presented by itself (Hebb, 1972). Their findings suggested to Müller and Pilzecker that memories are not instantly formed by experience. Instead, there must be a "perseverated" state in which memory items such as nonsense syllables are held and in which they can be disrupted, preventing eventual storage.

Over time, Müller and Pilzecker observed that the perseverated state is time limited: If there was a sufficient time period between the two lists of nonsense syllables, they would not interfere with each other. This pointed to a second, stable memory state in which memories would eventually be stored. Müller and Pilzecker named the process of transformation from a labile state to a stable state *Konsolidierung*, thus coining the term now used for the process we explore in this monograph. Müller and Pilzecker estimated the time course of *Konsolidierung* (i.e., consolidation) in humans to be of the order a few minutes, but this was specific to the list-learning experiments they were conducting. Later on, researchers would find that the consolidation period for other kinds of memory items can be considerably longer.

Modeling Memory, Take 1: Two Leaky Buckets

A simple yet workable model of memory consolidation is composed of a pair of leaky buckets (figure 2.1). Memory, represented by water, is poured into the first, leakier bucket. From there it leaks out quickly, much of it being lost, but some of it flows through the spigot into the second, less leaky bucket. The first, leakier bucket corresponds to the labile memory state; it is difficult to keep water in this bucket, and the volume of water in

Labile

Stable

Figure 2.1
Memory consolidation as a pair of leaky buckets. Memory, represented metaphorically as a "liquid," leaks out of both buckets, but it leaks much more rapidly from the first bucket than from the second bucket, which represent short-term (labile) and long-term (stable) memory, respectively. Some of the liquid "flows" through the spigot from the labile to the stable bucket. Although this liquid will ultimately be lost, it is retained for a much longer period than it would have been had it not flowed from the first to the second bucket. This "flow," or transfer from the labile to the stable store (bucket), is a metaphor for memory consolidation. (Drawing by co-author Patrick Watson.)

it diminishes rapidly. The second bucket corresponds to a more stable form of memory; water also leaks out of the second bucket, but the rate of leakage is much slower from the second than from the first bucket. Any water that flows into the second bucket is retained for a much longer period.

If the combined volume of water in the two buckets is plotted (solid curve in figure 2.2), it resembles experimental memory (or, more precisely, forgetting) curves for normal human subjects. An example of human memory (or forgetting) data is shown in figure 2.3. The data (shown as dots) were collected by psychologist Harry Bahrick. Over a period of decades, he measured the memory of Spanish of a large group of Americans who had studied the language as high-school students but used it rarely thereafter (Bahrick, 1984). The data shown in figure 2.3 specifically quantify Spanish vocabulary retention. The solid curve in figure 2.2, which is reproduced in figure 2.3, is in qualitative agreement with this data, suggesting that the two-bucket model provides a reasonable first

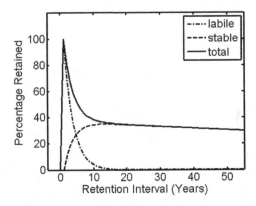

Figure 2.2

Graph of the contents of the two-bucket model. Initially, memory is held in the labile store (the first, very leaky bucket). It leaks out rapidly, but a small number is transferred (flows) from the labile store into the stable store (the second, less leaky bucket). After some time, the amount of memory held in the stable store exceeds that in the labile store. On a longer timescale, total memory is almost entirely a function of the amount of memory remaining in the stable store. On a very long timescale, memory fades from the stable store, and the total memory remaining approaches zero. The curves represent a simulation of the two-bucket model in which each time step represents 1 year, and in each year 30% of memory leaks out of the labile store, whereas only 0.4% leaks out of the stable store, and 11% of memory is transferred from the labile to the stable store. The total memory curve (solid line) agrees well with the human forgetting data plotted in figure 2.3.

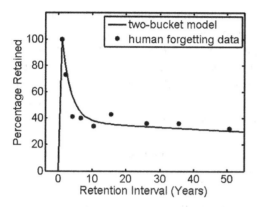

Figure 2.3
A human memory retention (i.e., forgetting) curve. The data are from Bahrick (1984), who measured the memory of the Spanish language of a group of 773 Americans who had studied Spanish as high-school students. The data points (dots) are the average of the Spanish vocabulary words each subject retained expressed as a percentage of the number each subject could remember as a high-school student. The data are consistent with the two-bucket model; the solid curve is the total memory curve reproduced from figure 2.2 (solid curve). The parameters of the two-bucket model (rates of loss from each bucket and rate of transfer from the first to the second bucket) were adjusted to fit this particular data set.

approximation to, and a mechanistic explanation for, the temporal properties of the human forgetting curve.

The human forgetting curve and the two-bucket model nicely illustrate the fact that, indeed, memory is gradually lost, even from long-term stores, in normal humans. Memory loss due to amnesia, which we will describe in a subsequent section, is measured relative to this background of normal memory loss, or normal forgetting, over time. The exact time course and rate of forgetting in the two-bucket model is dependent on the leakiness of the buckets and the rate of flow through the spigot, and these parameters can be adjusted to improve its fit to any data set. However, the point of presenting the two-bucket model is not to show how well it fits specific data sets but to use it as a starting point for discussions of other models that take more of the details of memory consolidation into account.

The First Neurobiological Explanation of Consolidation

A more detailed model of consolidation requires some approximation of what the neurobiological process actually entails. One of the first, neurobiologically plausible explanations for consolidation came from neuropsychologist Donald O. Hebb, who in 1949 postulated that recurrently connected sets of neurons represent labile memories in terms of neural activity and stable memories in terms of changes in connection strength (Hebb, 1949). At first, this "reverberating circuit" of self-reinforcing neural activity represents input from the senses as a dynamic electrochemical signal. Eventually, these reverberations "potentiate," or strengthen, the connections (synapses) between the firing neurons, such that future inputs (which could be thought of as cues or hints) can reinstate the reverberating pattern of neural firing. Through this process, a dynamic electrochemical signal is converted into static synaptic strength values.

Because neural firing is more easily disrupted than the strengths of the synaptic connections between neurons, reverberation and potentiation provide a plausible explanation for Müller and Pilzecker's findings. In addition to being plausible, Hebb's model is also extremely elegant because it is simple yet has great explanatory power. It shows how short-term memory could be dynamic while long-term memory could be static. It also shows how these two forms of memory could be instantiated by the same neural memory structure: short-term memory as dynamic reverberation in a recurrent loop and long-term memory as static potentiation of the synapses of the same neurons that compose the loop.

Hebb theorized that when the activity of one neuron leads to the activity of another, the synapse between them becomes stronger. Students of consolidation capture the essence of this *Hebbian learning* with the mnemonic: "neurons that fire together, wire together." Neurons so "wired" are the physical substrate necessary to produce stable memories. In other words, they are the stable physical (neurobiological) structures that support reconstructible memory representations. In 1979, neuroscientists Tim Bliss and Terje Lomo's discovery of long-term potentiation, a synaptic mechanism that strengthens the connection between two simultaneously active neurons, provided empirical evidence for Hebb's proposed mechanism of memory consolidation (Bliss & Lømo, 1973; Lømo, 2003).

Retrograde Amnesia and Gaps in Time

These neurophysiologic interpretations were acceptable to scientists working at the level of neurons and synapses, but they conflicted confusingly with parallel consolidation research in psychology. Müller and Pilzecker's theory of consolidation offered a new perspective on memory loss beyond simple forgetting, and psychological researchers realized almost immediately that disrupted consolidation provided an explanation for a specific form of amnesia known as retrograde amnesia (W. McDougall, 1901; Burnham, 1903). Retrograde amnesia along with its more common counterpart, anterograde amnesia, constitute *global amnesic syndrome*, which results from specific forms of brain damage (Markowitsch & Pritzel, 1985).

Amnesia refers to many types of memory impairment and may be induced by a variety of neurologic or psychologic pathologies (Markowitsch & Pritzel, 1985). However, damage to a particular cluster of forebrain structures reliably produces a dense and permanent global amnesic syndrome. The forebrain is the foremost part of the vertebrate nervous system (Butler & Hodos, 1996). Neuroanatomically, the forebrain is divided into the diencephalon and the telencephalon. The main diencephalic structure is the thalamus, and its major function is to relay input to the telencephalon from other brain regions. The main telencephalic structure is the cerebrum, which is divided lengthwise into two cerebral hemispheres (structures closer to or farther from the midline dividing the hemispheres are termed *medial* or *lateral*, respectively). What we commonly think of as "the brain," the cerebrum is further divided into frontal, occipital, parietal, and temporal lobes. The medial part of the temporal lobe is critical for memory formation.

The hippocampus, a seahorse-shaped structure that curls around within the medial temporal lobe (MTL), is arguably the most important memory structure in the brain because hippocampal damage can result in dense anterograde and retrograde amnesia. The amygdala and the entorhinal, perirhinal, and parahippocampal cortices are positioned near the hippocampus within the MTL and support its memory related functions. A trunk of nerve fibers called the fornix carries the main outflow from the hippocampus to the mammillary bodies, which in turn send nerve fibers to the thalamus. Together, the MTL, midline thalamic structures, and the mammillary bodies are critically involved in memory (figure 2.4), and damage to any of them can result in amnesia and deficits in memory consolidation (Markowitsch & Pritzel, 1985).

The most obvious and consistent symptom of MTL injuries is what is known as *anterograde amnesia*, an inability to make new memories. This condition prevents patients from consolidating or recalling any new experiences after the time of their injury. Most of these patients, however, also lose some past memories: they have some degree of *retrograde amnesia*. The memory impairment due to retrograde amnesia is most dense for the period immediately before the injury and becomes less dense as time recedes backward. Patients are often able to remember events from the

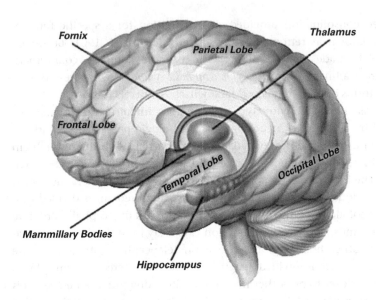

Figure 2.4

Brain areas associated with declarative memory disorders. The cerebral cortex is subdivided into the frontal, parietal, occipital, and temporal lobes. The temporal lobe contains a set of neural structures associated with memory disorders. Foremost among these is the hippocampus, but also important are the amygdala and the entorhinal, perirhinal, and parahippocampal cortices that are associated with the hippocampus. The fornix is a bundle of nerve fibers that carries a large portion of the outflow from the hippocampus to the mammillary bodies, which in turn send nerve fibers to the thalamus (particularly the anterior, or frontal, parts of it) (Butler & Hodos, 1996). Damage to any of these structures can result in dense anterograde amnesia (the inability to make new memories) and a limited retrograde amnesia, the temporal extent of which depends on the extent of the damage to these structures (see text). (This image is modified from one that is available in the public domain.)

distant past, such as childhood events, but their memory fades for events that occurred closer and closer to the time of the injury, until they reach a period of dense total amnesia for the injury itself and the time period just before it (Markowitsch & Pritzel, 1985). Amnesic syndromes vary greatly in their pathology (Nadel et al., 2003), but this pattern of temporally graded retrograde amnesia and dense anterograde amnesia is typical of MTL damage.

Already in 1881, Théodule Armand Ribot (1839–1916) described the retrograde amnesia curve (a more recent version appears in figure 2.5), in which more temporally distant memories are well preserved, whereas those

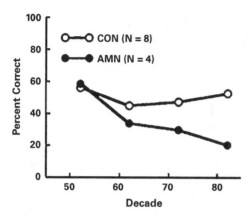

Figure 2.5
Typical retrograde amnesia data (Squire, 1992). Test subjects were amnesic patients
(AMN) with damage to the hippocampus and associated MTL structures (see figure
2.4) or age-matched, normal controls (CON). The data were gathered by assessing
how many "famous faces" each participant was able to identify from a corpus of
faces of individuals who became famous during a particular decade. The normal
(CON) curve displays the expected slow loss of long-term memory with time. The
increase in memory for the faces that became famous in the 1950s probably reflects
the "reminiscence bump" for these test subjects (amnesics and controls). The reminiscence bump is an increase in retention of the memories that people acquire
during their formative years (Conway et al., 2005). Amnesic patients and controls
were equal in performance for the oldest faces, those that became famous in the
1950s, but the patients had a distinct impairment on the more recent faces (1960s
through 1980s). The pattern of memory loss in which recent memory is lost but
remote (in time) memory is spared is characteristic of the retrograde amnesia that
results from damage to the hippocampus and related structures. (This image is available in the public domain. Source: Squire, L. R. 1992. Memory and the hippocampus:
A synthesis from findings with rats, monkeys, and humans. *Psychological Review, 99,*
195–231.)

from times closer to the amnesic event are lost (Ribot, 1881; Butters et al.,
1984). Not all studies suggest a smooth temporal gradient for retrograde
amnesia, but most find that damage to the hippocampal complex produces
a retrograde amnesia for recent memory but spares more remote memory
(Squire & Bayley, 2007). Figure 2.6 depicts this consensus view. In comparison with normal subjects, amnesic patients often have full memory for
distant (remote in time) events but reduced memory for events that
occurred recently with respect to the time of damage to their MTL or hippocampal complex.

Subject	Memory for remote events	Memory for recent events
Normal	Full	Full
Amnesic	Full	Reduced

Figure 2.6
Consensus view of the temporal limits of retrograde amnesia. Retrograde amnesia is loss of memory for events that occurred before the time of the damage or disruption that caused the amnesia. The terms *recent* and *remote* reference this time. The retrograde amnesia that results from damage to the hippocampus and related structures is characterized by loss of recent memory but sparing of remote memory.

In light of Müller and Pilzecker's data, it seemed logical to psychologists at the time (e.g., W. McDougall, 1901; Burnham, 1903) that a brain injury that results in both retrograde and anterograde amnesia has disrupted the process of transformation of labile memories to stable ones, as if the faucet between the first (more leaky) and second (less leaky) bucket had been plugged. Patients can retain stable memories from their already consolidated distant past, but the memories that were in process, stored in the labile state at the time of the injury and not yet consolidated, are disrupted by the insult, thus producing a period of retrograde amnesia. Because brain injury has interrupted the transformation process, no new long-term memories can be formed either, producing anterograde amnesia. The Müller and Pilzecker consolidation idea provided an explanation for the pattern of memory loss in retrograde amnesia, but there was a time-course discrepancy. Müller and Pilzecker's perseverated state lasted only seconds or minutes for their list-learning tasks, but retrograde amnesia patients are often missing memories for months, years, and even decades (see figure 2.5).

One such patient, who has become the canonical example of global amnesic syndrome, is Henry Gustav Molaison (1926–2008), known for privacy reasons only by his initials, H.M., until his recent death (Carey, 2008). In 1953, neurosurgeon William Beecher Scoville performed a bilateral resection of the MTL as a treatment for Molaison's uncontrollable epileptic seizures, which resulted from a head trauma Molaison suffered at the age of 9 years. After his surgery, H.M. had intact intelligence, personality, and short-term memory, but he was incapable of forming new long-term memories for facts and events (Scoville & Milner, 1957). H.M.'s retrograde amnesia extended for many years prior to his surgery, but he retained clear and vivid memory for facts and events from his childhood and teenage years. Henry Molaison, who underwent surgery for bilateral MTL resection at the age of 27, was left with intact autobiographical memory from before the age of 17 only (Corkin, 1984).

So profound was H.M.'s anterograde amnesia that, each time he met doctors or nurses who had worked with him for years in his assisted living facility, he would introduce himself as if to an entirely new person and claim to have never met them before in his life. He would re-read the same magazine article over and over, as if he had never seen it before. While dining, if the first plate was cleared away and a few moments allowed to pass, H.M. would eat from a second plate of food and could consume multiple meals without realizing it. Though he could learn new skills such as tracing a star in a mirror or reading reflected text, all new factual and autobiographical knowledge faded from H.M.'s mind and left no lasting impression.

One oft-repeated (and possibly apocryphal) anecdote of H.M.'s impairment happened "by accident." Hans-Lukas Teuber, a memory researcher studying H.M., was driving H.M. from his assisted living facility to the Massachusetts Institute of Technology (MIT). There was heavy rain at the time, and on the road in front of the two, a car spun out of control and went into a ditch. There was much excitement, and when Dr. Teuber had ascertained that the passengers of the wrecked car were all right and the car had been towed out of the ditch, he returned to his car, and he and H.M. continued their drive to MIT. At first they talked about the accident, but soon the conversation ended, and Teuber decided to test H.M. to see if he had forgotten the incident. "Why am I wet?" asked Teuber, who had been soaked by the rain during the incident. "You got out," replied H.M.— and then, after a pause—"to inquire about the way." Even something as vivid as a car accident faded from his mind: it could not be consolidated. Very little changed for H.M. after his surgery in 1953, but not because he lacked either labile or stable memories; what had been unknowingly destroyed was his ability to consolidate labile memories for facts and events into a stable state (Corkin, 2002).

In animal experiments, retrograde amnesia can be studied more directly than in human patients, but such research has raised more questions than its originators first sought to answer. Retrograde amnesia can be induced in animals by treatments such as hypothermia, electroconvulsive shock, protein synthesis–inhibiting drugs, and surgical brain lesions. By administering one of these treatments to an animal after training it to perform some task, neuroscientists can determine if a memory of the training has been successfully consolidated or if the training was still in a perseverated state when the treatment was administered. Two issues such research has raised involve the time course of the consolidation process, discussed next, and the differentiation of memory into types, addressed at the close of this chapter.

Just as there was a vast gulf between the time courses of consolidation for Müller and Pilzecker's list-learning experiments and H.M.'s retrograde amnesia, animal studies of consolidation in the mid-twentieth century also produced conflicting time-course information (Duncan, 1949; McGaugh, 2000). Even very similar experimental paradigms could produce consolidation time courses that differed by days or weeks. Though a perseverated state seemed to explain which memories would be lost in retrograde amnesia, the experimental data were still confusing. How long did certain memories perseverate? Why would different disruptions cause different amounts of retrograde amnesia?

Modulation and Systems-Level Consolidation

Two branches of literature arose from the confusion to explain the different findings on the time course of consolidation. First was American neurobiologist James McGaugh's work on memory modulation. McGaugh discovered that if a mouse is given a mild dose of strychnine immediately before a behavioral modification task, the mouse will be able to consolidate the memory of the task more quickly than a control mouse. Before McGaugh's 1972 work, it was thought to be impossible to accelerate the consolidation process, only to interrupt it. McGaugh showed that many factors modulate consolidation: it can be accelerated or retarded, not just disrupted (McGaugh, 1972; McGaugh & Herz, 1972). In light of these and earlier findings, McGaugh proposed that the time course of memory consolidation is not fixed but is affected by environmental or physiologic factors, and he termed this malleability "memory modulation" (McGaugh et al., 1979). Because so many contributing factors alter the duration and character of the consolidation process, it is counterproductive to try to "nail down" the exact time course of consolidation. McGaugh's work explains how different consolidation experiments should create conflicting time-course data by involving different sets of relevant factors.

The second insightful branch of literature, originating with American neuropsychologists Larry Squire and Neal Cohen (Squire & Cohen, 1979;

Cohen & Squire, 1980, 1981; Zola-Morgan et al., 1983; Squire et al., 1984), describes a different measure of memory consolidation's time course. Squire and Cohen extensively studied patients with MTL damage that had caused both retrograde and anterograde amnesia. The two researchers devised a creative test specifically to determine the exact extent and character of the patients' retrograde amnesia. They quizzed amnesic patients and nonamnesic controls on a series of celebrities' pictures, so-called "famous faces." Some of the faces were of people who had risen to prominence after the patients experienced MTL damage—they would not be expected to recognize these faces because of anterograde amnesia. Other faces were of individuals who were much in the news during past decades. Presumably, the patients would have been exposed to those faces because they were alive and active during those decades. The experimenters avoided iconic faces (e.g., Abraham Lincoln, Elvis Presley, etc.), because they are always current, and focused more on personalities who were famous over more circumscribed periods (e.g., Gary Cooper, Senator Joseph McCarthy, etc.). Because some amount of forgetting is normal, the performance of the amnesic patients on the famous faces task was expressed relative to that of the normal, age-matched control subjects.

As expected, control subjects had the best recall for the most recently famous faces, and their recall fell off gradually as the era in which the personalities were famous receded further into the past. In comparison with control subjects, the recall of amnesic patients was best for the most temporally distant faces. Further study indicated that the more extensive the MTL damage, the more extensive the amnesia, but in all cases the memory deficit for the amnesic patients was greatest for personalities that were famous during the decade preceding the MTL damage that caused the patients' amnesia. These data demonstrated that retrograde amnesia often extends for years before the date of the injury, suggesting that the time course for long-term memory consolidation must be of the order years or even decades.

To explain the huge discrepancy between the seconds-to-minutes time course of consolidation that Müller and Pilzecker described and the multiyear consolidation discovered in amnesics, Squire and Cohen suggested that consolidation could occur on at least two different levels in the brain. The theories of Hebb and the data of Bliss and Lomo (among others) support the idea of a "synaptic-level" consolidation, driven mostly by cellular biochemistry and influenced by modulatory factors like those McGaugh identified. This "low-level" consolidation occurs on the level of neurons and synapses where, through the previously mentioned process of potentiation, cellular mechanisms transform more transitory neural activity into more durable changes in the strengths of synaptic connections.

Distinct from this is "systems-level" consolidation in which various brain regions engage in a "reorganization" of memory that produces a long-term representation from items held in a more easily manipulated fashion in short-term stores. Squire and Cohen suggested that this higher-level form of consolidation would be influenced by existing knowledge. In systems-level consolidation, cognitive processes act on memory representations, and the neural structures responsible for this processing mainly include those of the MTL and the outer layers of the cerebrum, which are collectively known as the neocortex (Butler & Hodos, 1996). According to Squire and Cohen, the MTL holds and otherwise keeps track of short-term memories that are acquired quickly but stored in a more disruptable state and also organizes the formation of a less disruptable, long-term store in the neocortex. This process of consolidation is systems level in that it involves an interaction between various brain regions, and it can take much more time than synapse-level consolidation, presumably as much as years or even decades.

These two ideas—synapse-level and systems-level consolidation—explain the variability of consolidation's time course. Each type of consolidation is associated with its own variables that affect the duration of the process. The study of synaptic plasticity has advanced considerably in recent years using the new tools of molecular biology, but the specifics of synapse-level consolidation cannot be used to support the analogy we wish to make, on an abstract level, between individual and collective memory consolidation. We are instead concerned with systems-level consolidation and with the models created to explain, in more detail, the processes of reorganization of knowledge, specifically the integration of unconsolidated, short-term memories with already consolidated, long-term memories.

Modeling Memory, Take 2: Connectionist Models

Whereas the two-bucket model provides a useful shorthand for explaining the results of Müller and Pilzecker and of those who came later, including McGaugh, Squire, Cohen, and others, it ignores the specific content of memory. As a first step toward filling that deficit, investigators have used connectionist techniques to extend the descriptive power of memory models. The connectionist movement, which occurred in the 1980s and

1990s, strove to understand brain processes by simulating interconnected networks of neurons using computers. Computational neuroscientists still use *neural networks*, computer models that consist of neuron-like elements (often called *neural units*) that are interconnected with synapse-like connections of various strengths (usually called *weights*) (for a general introduction, see Anastasio, 2010). When the network is exposed to an input, some or all of the units respond. Unit responses depend mainly on the weights of the connections between the units. Essentially, each unit adds up its weighted inputs from the other units to form its own response. Learning rules can be used to modify the connection weights, thus changing how the network behaves over time, as if it were a sentient, adapting organism interacting with its environment. It is their reliance on the connections between neural units that gives connectionist models their name.

Pablo Alvarez and Larry Squire translated Squire and Cohen's conceptual model of memory consolidation into a neural network (Alvarez & Squire, 1994; Squire & Alvarez, 1995). The Squire and Alvarez model has two neural network components, one representing the MTL (including the hippocampus) and the other representing the neocortex. A memory item is stored in both places in this model: A strong, concentrated trace is stored in the MTL component, and a weak, diffuse trace is stored in the neocortical component. Using Hebb's postulate about neural learning, when the MTL representation is activated, it activates the units responsible for the diffuse neocortical representation. Because all of those units are firing together, regardless of their location in the neocortical component, they all wire together. Given enough time, the neurons in the neocortical representation will be so strongly interconnected that they will all activate each other whenever a memory cue is received (such as a partial memory) without needing to be driven or even primed by the MTL component. This model explains not only how the neocortex comes to represent long-term memories but also why the MTL seems to be critical, at least temporarily, in memory formation.

Neuroscientists James L. McClelland, Bruce McNaughton, and Randall C. O'Reilly took a more deeply connectionist approach in 1995. Their mechanistic model also uses two neural network components but they have radically different operational modes and need to work in complementary fashion to overcome each other's weaknesses. The interacting networks simulate systems-level consolidation by essentially shifting memory items from one kind of network to the other. The two network types are the Hopfield network, which can quickly grasp specific patterns, and the multilayer perceptron trained by backpropagation, a network that learns broad associations between patterns by slowly correcting errors it makes. These two network types are among the most familiar types of connectionist models (see, e.g., Anastasio, 2010).

Physicist John Hopfield invented the network that bears his name in the early 1980s (Hopfield, 1982). It consists of a single layer of interconnected neural units that feed back on each other (but not on themselves—self-connections are forbidden). It can learn a pattern in a single exposure by adjusting its feedback connection weights in proportion to the correlations of the elements of the pattern with each other. Let us unpack that a little.

A "pattern" is just a particular arrangement of ones and zeros, representing active and inactive units. This pattern might encode an item stored in short-term memory. When a Hopfield network has "learned" a pattern, it has developed the ability to make the activities of its units conform to that pattern, thereby representing a particular memory item. If the connection weights are proportional to the correlations of the elements of the pattern with each other, then each unit will develop some positively weighted and some negatively weighted feedback connections onto the other units. By dint of the positive connections, each unit that should be active for a pattern can help activate every other unit that should be active, and by dint of negative connections, each unit that should be active for a given pattern can also help *in*activate all units that should be inactive for that pattern. These interactions eventually (usually rather quickly) cause the activities of the units in a Hopfield network to re-form a learned pattern in a process that simulates a type of "remembering" as it might occur in certain parts of the brain.

For several reasons, the Hopfield network is a convenient model of the MTL (Rolls, 2007). First, a Hopfield network learns fast—it can assimilate a memory from a single exposure as the MTL is thought to do. Second, a trained Hopfield network will reproduce a whole pattern given an incomplete or corrupted version of it. For instance, given a partial pattern that represents a "thumb," the network will complete the pattern and reconstruct the original item "hand." Third, due to its feedback connections and dynamics, the Hopfield network can rehearse a pattern over and over. So if a Hopfield network had been trained to recognize a whole hand, it will, given a cue that leads to recall of "hand," simply repeat "hand" (i.e., "hand" has become a stable network state). The benefit conferred by the ability of a Hopfield network to learn fast and to reassemble and rehearse patterns verbatim is balanced by the cost of its limited capacity: The number of patterns a Hopfield network can reliably learn is a small fraction of the number of units in the network (Hopfield, 1982). This limitation can be overcome when the Hopfield network cooperates with another network type.

Whereas the Hopfield network is an autoassociator, which associates the same pattern with itself, a multilayer perceptron (Rumelhart & McClelland, 1986) is a heteroassociator, which associates two different patterns together. Having been trained with associated, or paired, patterns, the multilayer perceptron will produce a desired pattern of output activity given the corresponding pattern of input activity. This also requires a bit of unpacking.

A perceptron (Rosenblatt, 1958; 1958/1983), in contrast to a Hopfield network, is a neural network that does not have feedback connections. Because feedback is not allowed, a perceptron needs two or more layers of units, so that one set of units can send feed-forward connections to another set of units. To do anything interesting, a multilayer perceptron needs three or more layers: The first layer is the input layer, the last layer is the output layer, and the intervening layers are called hidden layers (Rumelhart & McClelland, 1986). Together, the multiple, feed-forward layers of units in a multilayer perceptron can accomplish extraordinarily brain-like forms of information processing (McClelland & Rogers, 2003).

Although the multilayer perceptron is capable of almost any input–output association (or transformation), it achieves its greatest effectiveness when used as a classifier, or categorizer. For instance, when given "hand," "foot," "arm," or "head" as inputs, it can produce "body part" as output, and when given "trout," "salmon," "halibut," or "flounder" as inputs, it can produce "fish" as output. Its ability to generalize gives it an almost limitless capacity, because it can associate very large numbers of inputs with a much smaller number of output categories. This ability to generalize is one feature that makes the multilayer perceptron an appropriate model for parts of the neocortex. Another is the manner in which the multilayer perceptron acquires its ability to associate patterns and to generalize.

Like many parts of the neocortex, a multilayer perceptron is not "born" with the ability to associate patterns and to generalize but must learn to do so. Although the precise mechanism by which the real neocortex learns is still unknown, the multilayer perceptron learns by means of a neural network learning algorithm called backpropagation. This algorithm essentially finds the error between the actual and desired output of the network and "propagates" the error "back" to the hidden layers. If the network makes an error, such as giving the output "fish" in response to the input "hand," backpropagation reduces the error by changing the weights of the feed-forward connections between units in the various layers. The adjustment is slow. Many presentations of the paired associates (such as "trout" and "fish") are required for the multilayer perceptron to learn them successfully, but slow learning is another feature that the multilayer perceptron shares with the neocortex. For example, a child seems to learn quickly to shout "dog" at the appearance of a dog of any arbitrary breed, but it actually took many rehearsals of observing dogs and of pairing them with category "dog" before he was able to make this generalization reliably. In the computer, a Hopfield network can rehearse patterns such as "beagle" and "dog" and present them repeatedly to the perceptron. A multilayer perceptron therefore works well with a Hopfield network because the Hopfield network can rehearse verbatim inputs and desired outputs long enough for the multilayer perceptron to learn them.

There is another benefit to a Hopfield/perceptron pairing. A few years after Jay McClelland and David Rumelhart's research group popularized the multilayer perceptron trained with backpropagation as a model of learning in the neocortex, psychologists Michael McCloskey and Neal Cohen (1989) uncovered a major weakness in it: "catastrophic interference." When the network is sequentially trained (i.e., it learns one set of input–output pairs and then a second set), the second set of associations almost completely overwrites or "catastrophically interferes" with the first set. Catastrophic interference does not occur in neurobiological learning, as everyday experience attests—learning, for example, to classify mushrooms as edible or poisonous does not wipe out all other previously learned classifications. Clearly, the multilayer perceptron trained with backpropagation cannot stand alone as a model of learning and memory formation.

A major impetus for McClelland and his co-authors in their development of the two-element consolidation model was to provide a remedy for catastrophic interference: "interleaved learning." In this scenario, input–output patterns from the old training set are alternated ("interleaved") with those of the new training set. In interleaved learning, the network is continually trained on a superset of all the paired associates it is required to learn, new and old. This remedy requires a memory buffer that can store the new patterns as well as the old, and one that can re-present them to the multilayer perceptron over many repeated iterations so the perceptron can generalize over them. Thus McClelland, McNaughton, and O'Reilly (1995) combined a Hopfield network with a multilayer perceptron to exploit the strengths of a system capable of both fast and slow learning (cf. Smith & DeCoster, 2000).

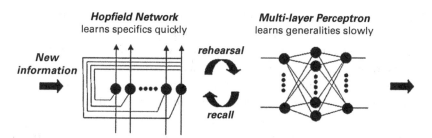

Figure 2.7

The two-box neural network model of memory consolidation proposed by McClelland et al. (1995) is composed of a Hopfield network and a multilayer perceptron. The Hopfield network consists of a single layer of neural elements (or units) that feed back on each other, and the multilayer perceptron contains at least three layers of units: input, hidden, and output. The multilayer perceptron has no feedback connections; instead, the input units project to the hidden units and the hidden units project to the output units. Memory items are represented in these networks as patterns of "activity" of the neural units, which are numerical values that indicate the responses of the units. The Hopfield network is able to learn quickly and retain verbatim patterns, but its capacity is limited. In contrast, the multilayer perceptron is able to extract slowly the structure inherent in a set of input and output patterns, and its capacity is essentially unlimited. In the model, the Hopfield network can repeatedly present patterns to the multilayer perceptron and thereby "train" the perceptron. The patterns stored by the single layer of units in the Hopfield network could be divided in two, with the first and second halves corresponding respectively to the input and output training patterns. The multilayer perceptron can also download patterns to the Hopfield network, so that old patterns can be relearned along with new patterns. This prevents the perceptron from forgetting old patterns, and it allows old patterns to be integrated with new patterns in the ongoing process of memory consolidation.

Figure 2.7 schematizes McClelland and co-workers' (1995) dual-process (two-box), complementary, memory consolidation model. The Hopfield network (on the left) models fast learning and labile (verbatim but limited) memory, and the multilayer perceptron (on the right) models slow learning and stable (generalized but practically limitless) memory. The McClelland model therefore accounts for some of the mechanisms that might underlie systems-level memory consolidation, but as we will explain in chapter 4, it leaves unaccounted several others.

We will often refer to the model of McClelland and co-workers (1995) as a "two-box" model because it has two major components. For the same reason, we could also refer to the two-bucket model as a two-box model.

Later (in chapter 4) we will introduce our own four-box (or, more precisely, three-in-one box) model. In all cases, we think of "boxes" simply as place-holders, and we can imagine filling them with various processors including buffers, relaters, or generalizers or with memory structures such as rever-berating neural circuits, synaptic weights, newspapers, posters, books, monuments, or museums. Flexibility in thinking about the contents of boxes allows us to better explain the models we consider in this book.

The two boxes in the model of McClelland and co-workers (1995) contain neural networks (see figure 2.7). Neurobiologists often complain that neural network (connectionist) models oversimplify neurobiology. This is a valid complaint, but, to the extent that the models successfully reproduce observations, they can provide insights into phenomena that go beyond the experimental data on which they are based. They also serve as powerful conceptual tools. Both the Hopfield network and the multi-layer perceptron illustrate how a collection of interconnected units can act synergistically to produce behavior of which the unconnected collection would be incapable. Both are intended to simulate aspects of individual memory, but with a little imagination we can also use them to provide insight into possible mechanisms of collective memory. As we will describe in the next chapter, there are basically two conceptions of collective memory: one that is merely the sum of the memories of the individuals who compose the group and another that is more than the sum. Contem-plation of the neural networks shown in figure 2.7 could lead one to imagine how a group of separate people could, if they interacted, store and retrieve memories that are greater in number and richer in detail than would be possible for a single person acting alone. We will retake this thread in the next chapter.

What Kinds of Memory Consolidate?

Finally, let us address the distinction between types of memory, because this distinction appears in both H.M.'s case and in animal consolidation research. It is important to realize, first of all, that authors have been dif-ferentiating between types of memory since at least Aristotle, as it is obvious to the careful observer that humans learn and remember various kinds of information (Herrmann & Chaffin, 1988). Findings on H.M. provided the first conclusive evidence that different brain regions mediate the formation of different kinds of memory, as resection of the MTL in H.M. produced profound anterograde and retrograde amnesia, but only for facts and events; H.M could learn and retain new skills as well as a normal person could

(Scoville & Milner, 1957). Subsequently, many different types of memory were distinguished. For instance, cognitive psychologist Bill Brewer has identified personal (or autobiographical) memory, semantic memory, generic perceptual memory, motor skill, cognitive skill, and rote linguistic skill (Brewer, 1986). Perceptual memory could be further separated into visual, auditory, haptic, gustatory, or olfactory memory, and so on.

Most memory scientists today make a major distinction between "declarative" and "procedural" memory. This distinction was originally based on the content of the memories. Declarative memory essentially comprises that which can be enunciated, or "declared," whereas procedural memory encompasses practiced abilities, or "procedures." A simple short-hand is that memory for facts and events is declarative, whereas memory for skills is procedural. A more precise way to differentiate between the two is whether an individual "knows that" or "knows how"—and although "knowing that" is considered to require consciousness, "knowing how" may be unconscious (Squire & Zola, 1996). Thus, knowing that your name is Horatio is declarative; knowing how to ride a bicycle is procedural. Because "conscious" and "unconscious" memory seems a bit difficult to come to grips with, other authors (e.g., Schacter, 1987) equate declarative with "explicit" memory and procedural with "implicit" memory.

Neuroscientist Endel Tulving has argued for a distinction within declarative memory: that between "episodic" and "semantic" memory (Tulving, 1972, 1983). Episodic memory, sometimes called "autobiographical memory," concerns events, or "episodes," that an entity experiences and that can be placed in the past. Semantic memory, by contrast, involves nonexperiential knowledge—"facts" in the general sense of the word. The difference between episodic and semantic memory is the difference between remembering the time you went apple picking and knowing (remembering) that apples grow on trees. Whether non-human animals possess episodic as well as semantic memory is still debated (e.g., Griffiths et al., 1999; Clayton et al., 2001), but this distinction is less significant for our discussion of memory consolidation than that between declarative and procedural memory.

An apparent difficulty involving animal studies of consolidation arises from the attempt to analogize between humans, whose memory has been divided according to what memories can be *declared*, or spoken, and other animals, who cannot "declare" their memories. Laboratory animals cannot tell researchers "I remember this maze" or "I know that receiving food means I did the task correctly," so, strictly speaking, they do not have declarative memory. Nevertheless, certain experimental designs enable laboratory animals to "declare" operationally. One such is the delayed-match-to-sample task in which an animal (typically a monkey) is shown an object and must then choose the same object from among distracter objects after a delay period, during which no objects are visible. If the delay period is sufficiently long, normal monkeys can choose correctly, but monkeys with MTL lesions (such as that which H.M. received) cannot (Mishkin, 1978). Because such tests are demonstrably hippocampal dependent, and because declarative memory consolidation requires the hippocampus in humans (Eichenbaum, 2000), experimentalists equate successful learning and completion of tasks such as the delayed-match-to-sample with consolidation of declarative knowledge of the task.

The distinction between declarative and procedural memory is a mixed blessing for our project. On the one hand, if collective memory exists, and we argue in the next chapter that it does, then a collective, through the individuals that compose it, should be able to "declare" its declarative aspects. Correspondingly, if a group of individuals can, through their interactions, consolidate and remember collective-declarative memories, then the "declarations" themselves should be readily observable. Part of the challenge for us is to distinguish individual-declarative from collective-declarative memory, and we argue in the next chapter that we can. Observability, combined with distinguishability between collective and individual forms, makes collective-declarative memory feasible to study. It is important to stress, however, that our focus is on memory *consolidation*, not on memory per se, whether individual or collective, and our goal is to establish an analogy between individual and collective memory consolidation. The relationship between the hippocampus, declarative memory, and consolidation is strong for individuals, and we hope to demonstrate that a connection between certain social segments, declarative knowledge, and consolidation is also possible on collective levels.

On the other hand, if collective memory exists, and again we maintain that it does, then it could have procedural as well as declarative aspects. Imagine, for example, the skills involved in sailing such large vessels that a single sailor could not possibly sail one of them alone. Obviously, the development of the art of large-vessel sailing required collective action over many generations and can be considered as an example of collective memory formation. Whereas certain aspects of large-vessel sailing can be declared, much of the actual, collective skill cannot be articulated, in the same way that an individual can describe bicycle riding but cannot, by language or diagrams or by any other means, impart to another the actual skill. This is not meant to imply that collective-procedural memory would be impossible to study but that it would be more difficult to study than collective-declarative memory.

In any case, it seems unlikely that any existing theory of declarative memory consolidation, including our own, can also satisfactorily explain procedural memory consolidation. One reason is anatomic. The brain structures responsible for consolidating procedural memory are different from those required for consolidating declarative memory. As mentioned earlier, although H.M. could not form new declarative memories because of the resection of his MTL, he *could* learn new procedural skills despite this lesion. This suggests that the brain structures responsible for forming procedural memories reside outside the MTL in individual humans. By analogy, to the extent that a society has collective-declarative memory and collective-procedural memory, they would be formed by different social structures.

Another reason is functional. The function of procedural memory is to preserve skilled ability, not to remember facts or form categories, and these different functions are acquired in different ways. Unlike declarative memory formation, some researchers suggest that the formation of procedural memory does not involve a consolidation from a labile to a stable state (e.g., Foster & Wilson, 2005). In chapter 6 we suggest what we believe is a critical difference between declarative and procedural memory formation. For now we simply restate that our exploration of individual and collective memory formation concerns the consolidation of declarative memory.

Conclusion

The study of memory consolidation has been a story of deepening understanding. Ebbinghaus and many thinkers before him observed that memories seemed to flow instantly and whole from perception into storage. Müller and Pilzecker found this assumption unsupportable and posited that memories held in a short-term state were consolidated, over a set time course, into a more long-term form. McGaugh discovered how to manipulate that time course. Squire, Cohen, and Alvarez described the process on a systems level, and McClelland, McNaughton, and O'Reilly produced a more detailed, mechanistic model of consolidation.

The driving force behind this progression of consolidation models has been new data, often gathered by researchers who had not set out to investigate memory consolidation per se. Müller and Pilzecker were studying list learning. McGaugh was studying non-human animal learning. Squire and Cohen were interested in amnesia. These threads of research data from neuroscience and psychology were woven into an understanding of memory consolidation that has been expressed in conceptual models with great explanatory power. We aspire to weave an even more extensive tapestry. We intend to gather data not only from the sciences but also from a heretofore untapped pool, the humanities. The analogical reasoning outlined in chapter 1 gives us a new tool with which to reason about memory interdisciplinarily. Like Marco Polo returning with spices from the East to enrich Renaissance cuisine, we hope that the new tastes from the humanities can enrich scientific studies of consolidation.

3   Defining Collective Memory

Although much remains to be understood about individual memory, no one really doubts its existence; people know that individual memory exists from their own, first-hand experience. This seems not to be the case for collective memory. This chapter defines and defends the concept of collective memory, and it engages with several recent critiques of collective memory. It does not argue that collective memory is analogous to individual memory; we make our analogy between individual and collective memory *consolidation*. But we must establish that collective memory exists before we can explore its formation. It is valid to suppose that the same process can be applied in the formation of two different things, just as essentially the same process of baking can be applied in the preparation of a chocolate soufflé or a tray of chocolate-chip cookies. We maintain that the processes that form individual and collective memories are the same, even if the results are rather different on different levels.

We will strictly limit ourselves to *memory*, and specifically to declarative memory, as described in chapter 2. Just as computers have memories but do not necessarily also have brains or minds, our argument that collectives have memories is not intended to support any theory of a "collective mind" or "global brain," as these phrases call to mind totalitarian Newspeak (Orwell, 1949/1961) or science-fiction–worthy giant, pulsating brains (L'Engle, 1962). Motivation for collective action can emerge in groups (e.g., mob rule; Le Bon, 1895/1960), even though individual members of groups can, and generally do, retain their own "agency." Most scientists have no problem with the construct "mind is what brain does," such that mind and brain may coincide in individuals. For humanists working in today's poststructuralist paradigms, however, structures and agents do not necessarily coincide in collectives. Even if a particular collective shares a single narrative or worldview, the individual members of that collective can and will interpret and apply that framework within their own, multifaceted contexts.

This chapter returns to the discussion of metaphors and analogies begun in chapter 1. Based on a survey of the scholarly literature, it then provides a history of collective memory and grapples with the legacy of metaphor in scholarly understandings of it. We will argue that beneath the metaphors there is an actual phenomenon of collective memory, something that is bigger than individual memory. Specifically, we think of collective memory as more than the sum of the memories of the individuals that compose the collective. If collective memory exists then it must be formed, and then it can be used for the process of "collective remembering" that many humanists have described. Again, our defense will not prove that collective memory is analogous to individual memory (but it will not disprove it either). It will establish that collective memory exists as a phenomenon unto itself that is similar to individual memory but occurring on different levels and taking different forms. This chapter will also set the stage for an overview of our model of individual and collective memory consolidation that we present in chapter 4.

A History of Collective Memory (in) Metaphors

Collective and individual memory has, historically, been described in metaphorical terms. The practice of making metaphors for memory (as opposed to analogies) is also cross-cultural. Historian Douwe Draaisma's book *Metaphors of Memory* (2000), for instance, includes numerous examples, drawn from throughout human history, of people making metaphors to describe individual memory. Metaphors based on natural objects such as honeycombs, forests, or "Bologna stone" (phosphorescent barite ore) are just as common as those based on man-made objects such as wax tablets, palaces, and dovecotes.

One common, old, and obvious memory metaphor is that of storage. Comparative literary scholar John Hunter (2002) has described how this figure of speech and shaper-of-thought arose in the ancient Mediterranean. Before offshore bank accounts and Swiss deposit boxes, the best storage technology for both material wealth and a wealth of written knowledge was a treasury. The Homeric poems (ca. 700 BCE) use a "treasure cave" motif for the other, the magical, the feminine. Seven centuries later, Roman orators such as Cicero domesticated the Greek poet's rich but dangerous space of the cave, transforming it into the neat storeroom of knowledge a citizen needed to be successful in the disordered public sphere. Memory was once "like" a communal storage cave or treasure house. In more recent times many people have come to think of memory as being "stored" in a filing cabinet—or even more currently, on a flash-drive. So whereas facets of memory have been expressed using a number of different metaphors, these are not metaphors drawn solely from individual memory.

Even before anyone wrote about something called "collective memory," a number of theories suggested racial or cultural bases for collective identities constructed around common characteristics, pasts, and future goals. Many people in the nineteenth-century West believed that the ethnic or racial memory of a nation was passed down through heredity and was even rooted in the very ground: think *das Volk* (the people) and *Blut und Boden* (blood and soil). Such notions supported nationalisms that insisted on a collective identity based on historical inhabitation of particular lands; not just Germans but Czechs and Slovaks, Poles, inhabitants of the Occitan/ Languedoc regions of southern France, and other ethnic groups claimed unique identities for themselves based on supposed territorial possessions and dubious cultural distinctions (Hobsbawm, 1962, 1975; Geary, 2002).

Blut und Boden was an organic metaphor for what we might call German collective memory—organic in the biological or natural sense. American comparative literary scholar Laura Otis (1994) and neuropsychologists Daniel L. Schacter and Elaine Scarry (2001) have discussed this first explicit theory of collective memory—"organic memory"—a popular and quasi-scientific speculation in the late nineteenth and early twentieth centuries about the application of heredity and memory to groups. German physiologist Karl Ewald Konstantin Hering (1834–1918) first hypothesized "organic memory" in 1870, a full decade before Hermann Ebbinghaus began to test individual memory. Based on a Lamarckian understanding of the transmission of acquired characteristics, organic memory made analogies between heredity and memory and between individual and society. It claimed that individuals inherit in their blood, cells, or brains the traits and memory of their social group. French psychologist Théodule Armand Ribot (see chapter 2) even went to far as to write: "Heredity, indeed, is a specific memory: it is to the species what memory is to the individual" (Ribot, 1875/1973, p. 52).

During the decades of discussion of species and generations, but before the rediscovery of Mendelian genetics in 1900 (Bowler, 1989; Henig, 2000), organic memory with its pangenetic inheritance mechanism seemed to provide a stable explanation of collective identity. Although considered unscientific at the time, this theory nevertheless was espoused by the likes of German biologist Ernst Haeckel (1834–1919), American neo-Lamarckian biologist Henry Orr, and English amateur science writer and antievolution-ist Samuel Butler (1835–1902), who more famously authored *Erewhon* (1872), the satire on Victorian society. It turned out to be a dangerous idea: when taken to its illogical extreme by racial hygienists during the first half of the twentieth century, organic memory meant that identity is predeter-mined and unavailable to be shaped by education or cultural conditioning. Otis concludes that this blurring of individual and collective memory contributed to the Holocaust. Whereas once the Nazis considered Jews, Poles, and other "non-Aryans" re-educable, eventually they decided that the natures of these peoples were set and immutable, and so the only way to secure land and a future for "Aryans" was to displace, ghettoize, and finally kill these "undesirables."

If Laura Otis's conclusion is correct, then the first collective memory analogy, that of organic memory, was not only scientifically wrong but politically disastrous. Criticism of the more widely known analogy of society as organism (discussed in the next section) also contributes to the unease many scholars feel when confronted with any analogy between collectives and individuals. We raise three points to counter this justified misgiving. First, our analogy is restricted to memory consolidation, not to memory itself or to collectives more broadly. Second, we gather a vast array of scientific and historical facts in support of our analogy between indi-vidual and collective memory consolidation. Third, the political ramifica-tions of our analogy are hopeful rather than hateful (see chapter 12).

We hope to rehabilitate analogy as a conceptual tool in our study of individual and collective memory consolidation. The foregoing discussion indicates that both analogy and metaphor have been used to theorize col-lective memory, and that the former is more powerful (and potentially more dangerous) while the latter is limited. Still, metaphors are instructive not only in being like something else but in being unlike it. The treasure house metaphor, with which we began this section, bears this out.

The treasure house is an innocuous and limited memory metaphor. People *seem* to accumulate memories in the same way a treasure house stores treasure: memories are seemingly localized within the person and are added or taken away. In contrast, treasure houses store physical objects, whereas brains store intangible thoughts. Although one can say by way of metaphorical shorthand that individual brains "store" memories, one could not say that a storehouse "remembers" its treasure. The problem of the storehouse metaphor is even more apparent for collectives, because they use physical objects—cultural tools, or "treasures"—to aid their col-lective memory even more obviously than individuals do. But just as for individuals, the memories are in the discourses collectives create around objects, not in the objects themselves. It is reasonable to suppose that individuals cooperating in groups can more richly endow a treasure trove than could the same number of individuals contributing separately. Similarly, we could imagine that the discourses elicited in a group admiring their cooperatively acquired treasure would surpass the thoughts of a single individual privately contemplating his own most valued possessions (even, we might suppose, those of a very wealthy individual!). Thus, even the humble storehouse metaphor helps us establish that collective memory exists as more than the sum of its members' separate memories.

Maurice Halbwachs and Social Frameworks of Memory

Collective memory research tends to trace its origins to French sociologist Maurice Halbwachs (1877–1945), whose *The Social Frameworks of Memory* (1925/1992) serves as the ur-text for most humanists.[1] It was in the context of "organic" memory metaphors that Maurice Halbwachs and three of his intellectual foils wrote in the early twentieth century. Beginning in the 1890s, Halbwachs's mentor, French sociologist Émile Durkheim (1858–1917), found that existing theories about collectives did not explain the ways in which modern societies actually work. Traditional or premodern societies, according to Durkheim (1893/1997), function mechanistically: organized into self-sufficient productive units such as households and subsistence communities, individuals are integrated like cogs in a gearbox. They have stable roles and identities in a smoothly functioning social "machine." Conversely, modern societies work "organically": although the division of labor and the fragmentation of families alienates individuals from each other, a "collective consciousness" unifies them as a functioning whole. It was important to Durkheim that the modern individual be able to differentiate himself from society with a strong self-identity, so it is against this all-encompassing collective consciousness that individual consciousness positions itself.

1. Halbwachs most fully confronts "collective memory" in *Les Cadres Sociaux de la Mémoire* (*The Social Frameworks of Memory*, 1925). Most English-language readers encounter Halbwachs through *On Collective Memory*, Lewis A. Coser's 1992 abridged translation of *Les Cadres* and *La Topographie Legendairre des Evangiles en Saint-Terre* (*The Legendary Topography of the Gospels in the Holy Land*, 1941). Despite its obvious title, *La mémoire collective* (*Collective Memory*, 1950)—which was compiled by Jeanne Alexandre and published posthumously—is concerned more with the experience of time than with collective memory (Halbwachs, 1950/1980).

What is interesting is the way Durkheim inverts the common chronology of social machine and social organism. Usually, traditional societies are described organically, as in-touch with nature and with their bodies, whereas modern societies tend to be cast as impersonal and mechanistic. But it is modern, industrialized societies that Durkheim characterizes biologically, and relatively simpler, preindustrial societies that he describes in mechanical language. It is also worth noting that some scholars interpret Durkheim's organic metaphor metaphorically and others literally. The former allows us to understand society "like" a social "organism," whereas literal organicism yields a "collective mind," which we have already declined to consider.

In the 1920s, Halbwachs expanded Durkheim's collective consciousness into a theory about the way collectives integrate their pasts with their presents. Neither division of labor nor charisma but memory, he argues (Halbwachs & Coser, 1925/1992), is the basis for social cohesion. Moreover, it is an unfixed memory that each generation reinterprets according to its own perspective. Halbwachs extended this idea to a form of collective memory that he labeled "social frameworks," which shape individual memory as much as individuals reinterpret collective memory (if not more so). Thus, according to Halbwachs, individual and collective memory are inextricably intertwined.

A second influence on Halbwachs's theory of social frameworks was the French philosopher Henri-Louis Bergson (1859–1941). If Durkheimians tended to subsume the individual within a form of collective psychology, Bergson tended in the opposite direction, to the primacy of the individual. Most pertinent among Bergson's wide-ranging thoughts is his idea of "intuition" and the related concept of the "memory cone." According to Bergson (1896/1913), the way to connect with other people is by turning inward toward oneself; there one finds a spectrum of feeling that spreads outward, allowing one to intuit the feelings of others. The memory cone telescopes outward and backward from the plane of consciousness in the present to the unconscious past. It can be zoomed-in on specific images or memories or zoomed-out to bring generalities into view. Thus, what one knows about the past, especially one's own personal history, is a result of turning one's attention inward.

A third influence on Halbwachs's theory of social frameworks was Austrian neurologist-turned-psychiatrist Sigmund Freud (1856–1939), who also developed an internalized theory of memory (Hutton, 1993, 1994). Freud was a historicist who believed that a human being retains in his or her subconscious not only a complete record of his or her personal history but also a complete record of human history. Psychoanalysis is the tool Freud developed in his clinical practice to help patients analyze their conscious "screen memories" for clues to the real memories hidden behind them in the subconscious. With skill and effort, therefore, there was no memory that was unreachable. Later in his life, Freud (1913/1952, 1937/1939) believed that all of human history could, theoretically, be recovered from the human subconscious.

Although Halbwachs did not entirely neglect the individual, he strongly disagreed with the notion that memory is personal. Going to the other extreme, he insisted that, although it is individuals who remember, all of their memories are strongly socially mediated. "It is in society that people normally acquire their memories. It is also in society that they recall, recognize, and localize their memories," he writes (Halbwachs & Coser, 1925/1992, p. 38). Halbwachs derides the psychoanalytic view that memories are stored, whole, in an individual's self-consciousness when he scoffs,

There is no point in seeking where [memories] are preserved in my brain or in some nook of my mind to which I alone have access: for they are recalled to me externally, and the groups of which I am a part at any time give me the means to reconstruct them, upon condition, to be sure, that I turn toward them and adopt, at least for the moment, their way of thinking. But why should this not be so in all cases? (p. 38)

According to Halbwachs, we reconstruct our memories according to the attitudes and customs of the groups of which we are a part. "It is in this sense," writes Halbwachs, "that there exists a collective memory and social frameworks for memory; it is to the degree that our individual thought places itself in these frameworks and participates in this memory that it is capable of the act of recollection" (Halbwachs & Coser, 1925/1992, p. 38).

In these quoted passages, Halbwachs is specifically concerned with the relationship of individual to collective memory, the existence of neither of which he denies. Our emphasis in this monograph is quite different. Whereas Halbwachs focused on individual remembering within collective memory, we are concerned primarily with the processes of formation of individual and collective memory and the commonalities shared between them. Still, as memory scholars we recognize the debt we owe to Halbwachs. We agree with him that individual and collective memory both exist, and while we focus on the analogy between individual and collective memory consolidation, we acknowledge the two-way dependence of individual and collective memory on each other (Wertsch & Roediger, 2008; see also chapter 1).

In Halbwachs's view, individual memories are mediated by social frameworks that arise as part of collective memory. Individuals tap into social frameworks by remembering from the perspective of this or that social group. Indeed, according to Halbwachs, they could not do otherwise, so if we want to understand individual memory we should understand how these social frameworks of collective memory function. Mediation also explains how we can participate in collective remembering of events we did not personally experience or people we never knew, what Halbwachs called "historical memory." Because "we" identify with various social groups, and because a defining characteristic of collectives is a shared past that "we" could not have wholly created, the existence of the phenomenon of collective memory is self-evident.

To complicate matters, "It follows that there are as many collective memories as there are groups and institutions in a society" (Halbwachs & Coser, 1925/1992, p. 22). Thus, every group in a society possesses its own social frameworks that support its particular collective memories. Any one collective has an overarching group memory as well as many subdivided memories. For example, there is "American" collective memory as well as "Asian-American," "African-American," "European-American," and so on. We acknowledge the existence of multiple groups in any society and the possibility that they all possess their own collective memory.

Memory "In the Group" versus "Of the Group"

The next two sections briefly explore two contemporary scholars' applications of Halbwachs's theory of social frameworks. In his book *Voices of Collective Remembering* (2002), cultural anthropologist James Wertsch agrees with Halbwachs that society can influence individual memory. But Wertsch demurs at the notion of a fully fledged collective memory that exists as a phenomenon that transcends the combined individual memories of the individuals composing the collective. In contrast, sociologist Jeffrey Olick (1999) seems more willing to accept collective memory as a synergistic phenomenon that is more than the sum of individual memories. We support the synergistic viewpoint. Moreover, we suggest that Wertsch's viewpoint is compatible with ours, given a small change in perspective.

Wertsch is especially critical of the practice of using metaphors from individual memory when discussing collective memory. He does not dismiss memory metaphors out of hand, but he does think they bring with them all sorts of problems. One is that of a collective mind, which, as we have argued, is not a necessary concomitant of a collective memory.

Another is the strange tension between superficiality and unjustified (and unsupported) duplication of brain or neural processes on the social level. As we explained in chapter 1, we agree that making metaphors for collective memory from individual-memory terms is merely a colorful, if descriptive, maneuver. French historian Marc Bloch (1886–1944) warned Halbwachs against such *léger de main* in his 1925 review of Halbwachs's *Les Cadres* (Bloch, 1925). Our theory of collective consolidation is different because it draws from scholars' understandings of both individual and collective memory to make an analogy, not a metaphor (see also chapter 1).

Wertsch makes a distinction between memory "in the group," that is, of individuals' memories within some social group, and memory "of the group," a truly collective memory. Wertsch addresses these two concepts in his discussion of Frederic Bartlett's unique social psychology project in *Remembering* (1932). Bartlett actually cites Halbwachs with approval—as a fellow "psychologist," no less—for Halbwachs's extension of Durkheim's collective psychology into a theory about collective memory. Bartlett agrees with Halbwachs's explanations of how memory works in groups— how families, classes, and religions, for instance, influence individual remembering (Bartlett, 1932, pp. 294–296). What Halbwachs cannot do for Bartlett is settle the issue of the memory of groups themselves. Wertsch suggests that Bartlett rejects the concept of a strong version of collective memory, the notion that there is some kind of memory "of the group," because it is unprovable (Wertsch, 2002, pp. 21–22). For Wertsch, there can be no "strong" collective memory because there is no collective "head." This is equivalent to arguing that there can be no "strong," "of the group" collective memory because there is no collective mind, but we reject the notion of an obligatory interdependence between mind and memory. In his argument for "weak" collective memory—of individual memory situated "in the group"—Wertsch reinforces parts of Halbwachs's thesis, particularly those parts concerning social mediation of individual memory that we discussed above.

According to Wertsch, a collective memory is shared by the individuals in a collective, who each individually remember more or less the same version of the past of their collective. Because there is no collective "head" or "mind," the collective memory is distributed over the individual members in the collective. Wertsch explains "distributed collective memory" as everything members of a group can remember in interactions among "agents" and between agents and "cultural tools." Distribution among agents quite evidently resembles Halbwachs's conviction that individuals remember together, relying on each others' memories. Moreover, an "[individual] mind extends beyond the skin" (Wertsch, 1991, 1998) in a second way, with inanimate cultural tools such as books or other texts, online search engines, museums, statues, and historical sites and other *lieux de mémoire* (Nora, 1984, 1989). Wertsch focuses his analysis on the deployment of narrative as a cultural tool in Soviet and post-Soviet Russian remembering of the past, a case study to which we will return in chapter 12. Here we submit that the aspect of collective memory in which it is distributed among agents and cultural tools is completely consistent with the idea that the "strong" form of collective memory can emerge from a collective as a synergistic phenomenon and as more than the sum of individual memories.

The interactive nature of collective memory, as Wertsch describes it, reminds us of the "transactive" and synergistic group memory proposed by psychologist Daniel Wegner (1986). Sociologist Jeffrey Olick (1999) sees the same dynamics at work and makes a distinction similar to the one Wertsch makes but allows for the existence of both alternatives. According to Olick there are, on the one hand, aggregated or "collected" individual memories. Sociologists frequently use surveys, he notes, but these try to get at the whole by measuring its individual parts. You might think of "collect*ed* memory" as the sum of individual memories. On the other hand are the social factors, irreducible to individual actors, that more properly deserve the term "collect*ive* memory." They characterize collective memory as more than the sum of individual memories and include supra-individual forces such as crowd psychology or institutional inertia. "It is not just that we remember as members of groups," Olick writes, "but that we constitute those groups and their members simultaneously in the act (thus re-mem-ber-ing)" (1999, p. 342). We agree with Olick that group memory can be collect*ive* rather than merely collect*ed*.

We can draw on an analogy with the artificial neural networks we discussed in chapter 2 to show how Wertsch's (2002, 2009) notions of "strong" and "distributed" are compatible and can both be applied to collective memory. Neural networks are distributed systems composed of units (model neurons) and their weighted connections (model synapses). Through the interactions of their interconnected (and *distributed*) units, neural networks can store and recall patterns in ways that constitute idealized but definite (and *strong*) forms of memory. These abilities are network properties that would not emerge from a set of unconnected, isolated units. Likewise, it is easy to imagine that a group of interacting people could store and retrieve memories "of the group" that the same people, if left isolated and alone, could not.

Through their interactions, the members of a society generate cultural tools (i.e., collective memory structures) such as books, search engines, museums, social frameworks, and foundational narratives as aids to memory both "of the group" and "in the group." Because collective memory is distributed, many individuals contribute to the creation of cultural tools, and any one individual's experience of memory "of the group" and "in the group" is mediated—shaped and constrained—by and through these various objects and conventions. The two-way interaction between individuals and groups applies as well to subgroups and societies—each creates cultural tools (memory structures) that shape the experience of memory (and remembering) of the other.

Whereas we agree with Wertsch that collective memory is distributed, we disagree with his interpretation of Bartlett's position on collective memory. Bartlett—like his French colleagues—believed wholeheartedly in "the group [as] a psychological unit" (1932, p. 299), one with the ability to influence the thoughts and actions of its members. Upon closer reading, Bartlett rejects only the suggestion that, by analogy with individuals, groups must *necessarily* have memory over and above that of their separate members. Although there is no evidence from sociological studies to *prove* "group memory," he admits in 1932, neither is there any evidence to *disprove* it. Consequently, the possibility should not be dismissed. Bartlett indicates that if there had been evidence at the time from studies of groups as such, instead of only groups' influences on their members, he would have accepted it (Bartlett, 1932, pp. 293–300). He suggests, moreover, that it might have been only experimental design, and not the lack of the phenomenon itself, that prevented his fellow researchers from ascertaining the existence of collective memory "of the group." Seventy-five years later, we still lack hard experimental proof of collective memory "of the group," but we suspect it is only a matter of time before results appear from experiments demonstrating that collective memory is indeed more than the sum of the memories of the individuals who compose the collective. We will suggest a possible approach to such experiments later in this chapter.

Memory in Modern Societies

Our discussion of organic memory metaphors versus analogies comes together nicely with debates on memory "of the group" in a challenging quotation from the end of Peter Novick's 1999 study of the Holocaust in American memory. His work supports the idea of collective memory "of the group," even though Novick writes that he doubts collective memory is an appropriate concept for modern, technological societies:

When we speak of collective memory, we often forget that we're employing a metaphor—an organic metaphor—that makes an analogy between the memory of an individual and that of a community. The metaphor works best when we're speaking of an organic (traditional, stable, homogeneous) community in which consciousness, like social reality, changes slowly. . . . How appropriate the metaphor is for the very inorganic societies of the late twentieth century (fragmented rather than homogeneous, rapidly changing rather than stable, the principal modes of communication electronic rather than face to face) seems to be questionable. . . . We're a long way from traditional notions of the transmission of collective memories, let alone the sort that endure for centuries. (Novick, 1999, pp. 267–268)

Novick acknowledges an analogy between individual and collective memory but then refers to collective memory as an "organic" metaphor. He reverses Durkheim's inversion of mechanistic and organic societies to the nineteenth-century sense of slow-moving, organic "traditional" societies and fast-paced, fragmented "modern" societies. Collective and especially "long-lived" memories Novick associates with traditional societies; it is "questionable" whether modern societies are capable of producing anything of the sort. Yet Durkheim posited a collective memory because he saw modern European society as fragmented, as needing something to pull it together lest anomie result. For Durkheim, collective memory in modern societies was not only possible but necessary. From a more recent perspective, the existence of collective memories in fast and fragmented societies is not actually a difficulty, and for two reasons.

The first reason fast-paced, fragmented modern societies do not prove structural or functional hindrances to understanding collective memory is that memory *of* a collective depends on interaction among individuals, and the potential for interaction increases inexorably. Any tendency for communication technology to isolate individuals is more than counterbalanced by its ability to bring people together. Modern communication technology may decrease the importance of local, face-to-face interactions, but it increases communication within subgroups of all sizes (Wellman, 1999). Such enhanced interaction should facilitate the ability of subgroups to consolidate their own collective memories.

Second, Novick suggests that collective memory is slow to change and therefore not suitable for describing today's peripatetic collective phenomena. But James Wertsch (2002) points out that "Despite its claims for stability and constancy, then, it appears that one of the few genuinely constant attributes of collective memory is that it is likely to undergo change"

(p. 46). As psychologist James W. Pennebaker suggests, "Collective memories, then, are often cohort memories" (Pennebaker et al., 1997, p. viii). Thus, collective memories change with each generation, if not more often. This is the shifting of frameworks over time and generations that Halbwachs discussed (e.g., 1925/1992, p. 182), and which is consistent with our view that collective memory consolidation, like individual memory consolidation, is a continual process.

An Argument for Collective Memory

The solidifying of disciplinary boundaries in the late nineteenth and early twentieth centuries profoundly affected the way memory has been studied in the West. The reason the first half of Bartlett's *Remembering: A Study in Experimental and Social Psychology* was ignored when the book was published in 1932, and for another half century afterwards, was that experimental psychologists at the time worked within a behaviorist paradigm unconcerned with the subjective aspects of memory that are affected by internal states or social contexts (e.g., Watson, 1930). It took the cognitive revolution of the 1960s and 1970s for psychologists of memory to become interested in internal states (McGaugh, 1972; McGaugh & Herz, 1972). Only in the 1980s did psychologists rediscover the second half of *Remembering*. For their part, neuroscientists have recently begun to redraw the boundaries of their objects of study to integrate social effects: work on the neural bases of economic and moral behavior are two recent cases in point (e.g., Camerer et al., 2005; Moll et al., 2005).

A number of authors have complained that while investigators in the biological and psychological sciences have agreed on some basic principles of "memory" and identified various levels at which they study it (the synapse, neural networks, neuroanatomy, behavior, etc.) (Schacter, 1996; Roediger & Goff, 1998), in the humanities there has been little consensus as to what, exactly, sociologists, anthropologists, literary critics, or historians mean when they invoke "memory" or even whether they are talking about the same phenomenon as scientists at all (Wertsch, 2002, esp. pp. 32–34; also, Olick & Robbins, 1998; Winter & Sivan, 1999; and K.L. Klein, 2000, parses recent historians' use of "memory"). Some have expressed doubts that there even is such a thing as "collective" memory. If all memory is socially mediated, goes the argument, then what is the need of another concept above individual memory and besides history? This is not the assertion that French memoriologist Pierre Nora makes when he says, "We speak so much of memory because there is so little of it left" (Nora,

1989, p. 7). Nora does believe in something called "collective memory": His fear is that history has crowded it out of public consciousness. We will turn more fully to this discussion of individual memory versus collective memory versus history in chapter 12. What concerns us here is the relationship between individual and collective memory and the way in which the existence of the latter is revealed by the former.

American historian Susan Crane perceives an urgent need to "writ[e] the individual back into collective memory" (Crane, 1997). Crane argues (and we agree) that history and collective memory are not mutually exclusive; rather, the former is a subset of the latter (see chapter 12). She further wants to reinsert the remembering individual into collective memory, reasoning that it is individual people exercising their historical understanding that makes collective memory possible. This assertion seems to parallel historians Jay Winter and Emmanuel Sivan's proposition of a *homo agens* at the center of collective remembering: "He or she acts, not all the time, and not usually through instruction from on high, but as a participant in a social group constructed for the purpose of commemoration" (Winter & Sivan, 1999, p. 10). Their point is that collective memory is rendered observable through the externalization of it by individuals that are active in the process of collective remembering. Those who are so engaged are the proper site of collective memory research, argue Winter and Sivan.

Writing a few years before Winter and Sivan, Crane (1997) wishes to replace collective memory with socially contextualized individual memory. She observes,

[T]here is a body/body problem lurking in this [Halbwachs's social frameworks] theory of memory that is rarely alluded to: we all know that groups have no single brain in which to locate the memory function, but we persist in talking about memory as "collective," as if this remembering activity could be physically located. . . . None of this, however, addresses the fact that collective memory ultimately is located not in sites but in individuals. All narratives, all sites, all texts remain objects until they are "read" or referred to by individuals thinking historically. (Crane, 1997, p. 1381)

We agree with Crane that collective memory—memory "of the group"— cannot be located in a "single brain." We also agree that unless a human agent actively interprets or "reads" them, cultural tools such as monuments or books are just objects: A statue is not a memory, if by that we mean "a context-dependent reconstructed version of the past." But we conceive differently than Crane of the nature of collective memory and its "location."

Whereas the brain is the medium for individual memory, society is the medium for collective memory. Just as neurons in the brain, interacting through their synaptic connections, can store and recall patterns, so people in society, interacting directly and through cultural tools, can form memories and remember the past in ways they could not do as separated, isolated individuals. On both the individual and collective levels, we can distinguish between memory structures (e.g., synapses, neurons, books, museums, etc.), which are objects, and memory consolidation (memory structure formation) and remembering (via memory structures), which are processes. The processes of consolidation and remembering respectively form and use memory structures on the individual and collective levels.

Memory structures, as physical objects, can be physically located. Does that mean that the processes and the structures are co-localized? Well, yes, if we can say that the ticking of a wound-up clock is located in the clock, but then again no, because the ticking, but not the clock, vanishes when the clock winds down. Whereas a memory structure is material, a memory, or a remembrance, has no substance: It is pure dynamics. So we agree with Crane that remembered memories cannot be physically located; being context-dependent reconstructions, they vanish once the rememberer's attention is diverted. But the distinction between structure and process has little bearing on the question of whether collective memory exists.

A process cannot be localized in the same sense that the structure that supports it can be localized, but this does not imply that processes, compared with structures, are any less real. The point we are making here is that, just as the dynamic interaction between neurons in a brain produces individual memory, so the dynamic interaction between brains in a society produces collective memory. Moreover, this dynamic interaction on the collective level results in collective memory that transcends the merely combined memories of the individuals that compose the collective. As historians Winter and Sivan (1999, p. 6) themselves state concerning the nature of collective memory, "the whole is greater than the sum of the parts."

Some Informal Collective Memory Examples

We began this chapter by noting that no one really doubts the existence of individual memory, because people know it exists through their own personal experience. Perhaps, as we near the conclusion of this chapter, readers would be willing to consider that they also know that collective memory exists through their own experience. We offer three short vignettes illustrating that collective memory is at least different from, if not also more than, the sum of the memories of the individuals that compose the collective.

Consider this real event: Three adolescents commit an act of vandalism. The police officer called to the scene questions each of the three separately, not together. The officer would save time by questioning the three together, but he knows their memory of the event would be different if they remember it together. Consider next reminiscence at a family get-together. Each family member not only contributes facts that the others have forgotten but also triggers in other family members their own remembrances that they would not have had alone. The family memory is more than the sum of its parts. Finally, consider the justices of the U.S. Supreme Court, trying to remember the law that pertains to the case at hand. They rely not just on their individual memories for that area of law but also confer with each other and read the pertinent law from legal texts and consult past decisions to determine precedent. Clearly, each justice's memory for the pertinent law, in conference with each other and by referring to consolidated collective memory structures (law books and court records), is different, and likely more extensive, than it would be if each had remembered alone.

Perhaps readers who have not experienced exactly these situations can substitute experiences from their own lives as members of local communities, families, and professions. The point is that memory in any collective is not limited to the sum of its members' memories. Instead, collective remembering is a dynamic process involving interaction among its members, from which emerges something new that is different from the sum of the isolated memories of the individuals in the collective. The same holds for collective consolidation, a process involving an interaction among the members of a collective that produces stable memories that transcend those that a set of isolated, unconnected individuals could produce. Although individual and collective memory *consolidation* is the main topic of this monograph, we will consider, before we close this chapter, a suggestion concerning the way in which the existence of collective memory itself could be experimentally tested.

Outline of a Collective Memory Experiment

Knowing that introspection has never been a substitute for observable fact, we would like to offer another, informal collective memory example and then suggest the outline of an experiment that could be used to test whether collective memory is indeed more than the sum of separate, individual memories. Imagine a set of individual test subjects who each visit a staged, put-on haunted house alone. These subjects would all be members of the same social group, so the social frameworks that influence their experience and their memory of it are the same. Then, each individual has his or her own, individual memory of his or her own scary experience.

Now we bring this set of individuals together into a group, and they interact and relate their experiences. Some saw things others did not. Some remembered details better than others. Some remembered items that triggered recollection in others. Some were more or less scared by certain displays than others. Recalling their experiences in a group setting and hearing the recollections of others could possibly augment (by stimulating recall) and otherwise change (through comparison with others) the representation each has of his or her own experience.

An experiment could be designed around detailed accounts that the subjects write individually before they meet as a group and collectively after they meet as a group. Of course, we are comparing the memories for the same event before and after a group interaction but, as we will explain in detail in subsequent chapters, we see memory consolidation as a continual process anyway, so this would be a valid comparison. Before the subjects meet as a group, they are each instructed to write a detailed description of each room they visited in the (staged, put-on) haunted house and of each character in each room. They must also rate, on a scale of 1 to 10, how scared they were in each room. After the subjects meet as a group, they are instructed to write a detailed account according to the same format but to write it together, as a group.

The first step in data reduction is to combine all the individual accounts into a single account. The combined account would include all the details from every individual account. Redundancies would be omitted, but inconsistencies would be included (e.g., three subjects recalled that the bloody, severed head had brown hair, whereas two recalled that it had blonde hair). The scariness ratings for each room would be averaged. The next step is to compare the combined account with the group account on number of details, accuracy, and scariness ratings for each room. If the group account compared with the combined account is more detailed and more accurate, then that would provide a clear indication that the collective memory, in this case, is more than the sum of the combined, individual memories. It would also be of interest to see how the scariness ratings compare. Any differences would provide additional evidence that the collective memory is different from the combined, individual memories.

The exact experimental design and procedures would require some trial and error to work out, and the experiment would have to be repeated using many different sets of subjects and many different testing environments. Although each set of subjects should be from the same social group to ensure that each member has the same social frameworks, the experiment would need to be repeated in a number of different social groups. In addition to (or instead of) a (staged, put-on) haunted house, test environments could include visits to a zoo or aquarium, screenings of movies, and other attractions that the experimenters could thoroughly characterize in order to gauge properly the accuracy of the combined and group accounts. As for any other experimental approach in psychology, the development of experiments to test convincingly the existence of collective memory "of the group" would require vetting by competing laboratories. Although we strongly believe in the existence of memory "of the group," we obviously cannot predict the outcome of experiments to test it. We hope this chapter will inspire the kinds of experiments that can decide the issue one way or the other.

Conclusion

In this chapter, we have argued that collective memory exists as a synergistic, emergent phenomenon as opposed to a more or less common version of a group's past that is held in the individual memories of most of its members (e.g., Wertsch, 2009). What we like about the former is that it conforms better to our idea that collective memory consolidation is not just the collected consolidations of the group members but is a process analogous to individual consolidation that occurs on collective levels. In our view, the collective consolidation process produces collective memory structures (from books to myths) that subsequently evoke collective remembrances that transcend the individual level. But we admit that the idea of an emergent, transcendent collective memory is hard for an individual to grasp. Therefore, we point out that our model of consolidation applies equally well to the synergistic as to the separate but shared varieties of collective memory. Even if collective memory is merely separate but shared, it still requires that labile-shared memories are gathered, related, and generalized into stable-shared memories that most people in a group can more or less agree on and separately remember. Whatever form collective memory may actually take, it is important to reiterate that the subject of our book is not collective memory per se but individual and collective memory *consolidation*. The rest of this monograph will explain how our model applies to the consolidation of individual memory and to collective memories of various forms.

Although we stress that the focus of this monograph is on the *analogy* between individual and collective memory consolidation, we acknowledge the interaction between these levels. This interaction is critical for understanding the formation of many collective memory structures. Consider, for example, an autobiography. It is ostensibly the written remembrance of a single individual, but if memory is socially mediated and societal factors influence everything from the author's intended audience to her idiom, then the autobiography is to some extent a product of collective memory formation. Recent work by Berntsen and Bohn (2009) supports this view by showing how individual autobiographical memory is shaped by social norms and frameworks that specify the "plan for a life" that individuals in a group share.

In contrast to a written autobiography, consider a scientific textbook. Though a single author may have written it, the facts and theories it describes were obtained or conceived and, most importantly, vetted by a whole community of scientists. We will have more to say on this subject in the next chapter, where we will elaborate on how a scientific textbook is most definitely a product of collective memory consolidation.

Codes of law, like textbooks, are clearly also products of collective memory formation, but what about intangible yet equally influential products such as foundational myths and cultural narratives? If unwritten, then their "structure" consists of the brains of the individuals in the society who remember them and pass them on and together also change them along the way. It is easier to see how memory structures with more, well, *structure*, such as museums, monuments, and historical sites, are products of collective memory formation, because groups of interacting individuals are needed not only to build them but also to decide to do so (and typically the deciding takes longer than the building). Because individual memory is socially mediated, and because neurons and synapses can form part of the structure of collective memory, it becomes extremely difficult to draw a line that separates individual and collective memory structures. All the more reason, then, to suppose that individual and collective memories are formed through analogous processes. That they are is our main thesis.

4 Three-in-One Model of Memory Consolidation

Our overview in previous chapters of the literature on individual and collective memory has supplied background sufficient for us now to present an outline of our model of individual and collective memory formation. The outline we develop in this chapter will provide a framework for part II, where we will undertake a more extensive literature survey that will support our model. This chapter focuses on models of memory consolidation. In chapter 2, we alluded to a paradigm shift among scientists of memory consolidation from a "two-bucket" transfer to an ongoing process of reorganization. We described the strengths of the two-box model of James McClelland, Bruce McNaughton, and Randall O'Reilly (1995)—especially its multiple-systems approach—and began to explain why we believe this model no longer sufficiently captures what researchers have learned about memory consolidation in the intervening 15 years. At the same time, we fully acknowledge the influence that the model of McClelland and colleagues has had on our own theorizing. Here we describe the model of McClelland and colleagues at a level of detail appropriate to its importance in ongoing efforts to understand memory consolidation.

This chapter contains the nucleus of the argument we present in more detail in part II. The first section is devoted to previous models of memory consolidation and explains how those models led us to develop our own model. In the second section, we lay out our "three-in-one" model of memory consolidation. Whereas the two boxes in the model of McClelland and colleagues can be thought of as a "buffer," which holds verbatim memory items for the short term, and a "generalizer," which holds a compressed and efficient version of them for the long term, our model has a "four-box" configuration that includes the buffer and generalizer of McClelland and colleagues as well as two additional processors: a relater, which identifies associations between memory items, and an all-encompassing "entity." The entity is the whole "being," whether individual or group, that forms and uses memories for its own purposes. It is not itself an element of consolidation but rather subsumes the elements of the consolidation process, as well as all the nonmemory factors that influence it. After all, memory consolidation does not take place in a vacuum: The thoughts, feelings, goals, and desires of the remembering entity all shape the processes that form memories. Finally, the chapter closes with a re-reading of a familiar work, Thomas Kuhn's (1962/1996) *The Structure of Scientific Revolutions*, as an early and insightful case study of memory consolidation in a particular kind of collective: scientific communities.

Modeling Memory: The Two-Box Model

The "two-bucket" model is perhaps the simplest two-component (or two-box) model that is consistent with the reorganizational view that memory consolidation involves an interaction between separate neural structures implementing different memory functions. When a new memory item is received, the structures that can best represent this item for the short term (corresponding to the first, more leaky bucket) activate to store a trace of that item. The temporary but continual access afforded by short-term structures can extend this representation to other memory structures (corresponding to the second, less leaky bucket) that are more efficient at long-term storage. This process leads to a more durable representation of the original memory trace (Cohen & Squire, 1981; Schacter & Tulving, 1994). The memory essentially transitions from a labile state (the initial, temporary storage) to a stable state (the subsequent, more permanent representation). One could use the two-bucket model to approximate this change, but models based on neural networks provide greater insight into actual brain mechanisms.

In the mid-1990s, there was a surge of interest in consolidation models. For instance, in 1994, Pablo Alvarez and Larry Squire proposed a neural network model for this labile-to-stable transition that involved the transformation of memory representations from a less to a more distributed form (Alvarez & Squire, 1994). This model has two neural network components: one corresponds to the hippocampus (and MTL) and the other corresponds to the neocortex, and both store representations of items (see chapter 2). As in most neural network models, these representations amount to specific patterns of activity and connectivity among the units in the network: The patterns are stored and retrieved according to the interactions between the units, which depend on the strengths of the synaptic connections between them.

Initially, the hippocampal component stores a stronger and less distributed version of an item and the neocortical component a weaker but more distributed version. The units in each component are internally connected, and the components are connected to each other. Because its units initially have stronger connection weights, the hippocampal representation can be reactivated easily and—via its links to the neocortical component—activate the entire neocortical representation of the memory. Because all connections are Hebbian, over time the weights of the connections between the units in the neocortical component become strong enough that they can reinstate their own activity pattern by themselves. Activation of the neocortical network by the hippocampal network is no longer needed, and the short-term representation in the hippocampal component can fade away. In this way, a labile representation might be "stored" and made available for the short term in the hippocampal component, while the neocortical component gradually develops a stable representation of it for the long term. This model provided an explanation for the transition from a short-term to a long-term distributed store, but it did not address consolidation at the level of the individual memory item.

To explore this issue computationally, McClelland et al. (1995) developed a neural network model to account not only for how memory consolidates but also for how multiple, discrete memory items could be successfully stored in long-term memory even as they faded from short-term memory. Their main goal was to demonstrate the benefit of combining different memory systems to accomplish multi-item consolidation. As we described in chapter 2, this two-box model (figure 4.1) contains two components: one that stores and rehearses verbatim patterns (memory items), and another that exploits that rehearsal to produce a generalized and efficient representation of them. From the original consolidation theory's standpoint, the first "storage" or "buffer" box represents the labile, disruptable memories that are held in the perseverated state theorized by Müller and Pilzecker (1900). The second box represents the stable state into which memories are eventually transformed by the process of consolidation. Importantly, McClelland and colleagues had begun to develop a more sophisticated, reorganizational view of consolidation, as the first box is "teaching" the second to extract the rules by which a more generalized and efficient representation of the original patterns can be generated.

There is reason to believe that collective memory similarly possesses labile and stable memory states. Kansteiner (2002) points out that, of all

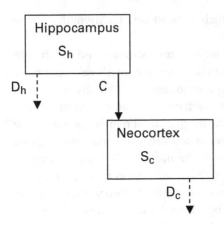

Figure 4.1
Schematic of the two-box model of memory consolidation proposed by McClelland et al. (1995). A memory trace in the hippocampus, S_h, strengthens the neocortical trace, S_c, at some rate C. The hippocampal and neocortical traces decay at rates D_h and D_c, respectively. This two-box model is equivalent to the two-bucket model shown in figure 2.1 and is a simplification of the two-network model shown in figure 2.7. (Adapted from McClelland et al., 1995, with permission of the American Psychological Association.)

persons, objects, and events that constitute "potential" (i.e., labile) collective memories, only a selected few become "successful" (i.e., stable) collective memories. Once we accept that there are these two aspects of collective memory—labile and stable—it remains to describe a system capable of transforming one into the other that represents the collective memory consolidation process. This we will do in greater detail in part II. For now it suffices to consider that collective, like individual, memory is not a monolithic process but rather one that depends on multiple, interacting components. There must be institutions that manage collective short-term memory (the way the hippocampus does in individuals), various kinds of societal "teachers," and finally stable representations of collective memories like the stable representations found in the neocortex. In both individual and collective entities, that which does not successfully consolidate is effectively forgotten.

The "storage" element that McClelland and colleagues used is a "stable attractor network" such as the one John Hopfield designed (Hopfield, 1982). As described in chapter 2, Hopfield networks excel at quickly and accurately learning the patterns presented to them. The network uses Hebbian covariational learning rules, in which the perennial tenet "neurons that fire together, wire together" is augmented by rules in which neurons that stay quiet together also develop mutually excitatory connections, and neurons that should have opposite activity states develop mutually inhibitory connections. The units in a Hopfield network have bounded states (from zero to one), and the connections between the units form positive and negative feedback loops by which they dynamically drive each other to the extremes of their activity (zero or one). Thus, a Hopfield network is a dynamic system that begins from some state (pattern of activation over the units) and "settles" from there to a stable state that corresponds to a remembered state.

A good visual analogy for this process is a bead of water rolling down the side of a cold glass. If an initial input (think of it as a hint or clue) matches part of a pattern the network has already learned, it can fill in the blanks and reproduce the remembered pattern, just as the smooth channel left by a previous drop guides a new drop down its path. The initial input settles into a stable pattern of activity in the network, which persists until the network receives a new input. The learning process of a Hopfield network is also quick in that it learns a pattern in one exposure. Its capacity is strictly limited by its number of units, but within its capacity it can recall a pattern verbatim and reiterate that pattern until it receives a different initial input and settles to a different pattern. Of course, all of the connections can be reset (zeroed), and the Hopfield attractor network can be retrained on a new set of patterns within its capacity.

Hopfield networks excel at quick storage and verbatim retrieval, but this speed and accuracy comes with drawbacks, because Hopfield networks learn specifics, not general features. If you showed a Hopfield network a poodle and a Pomeranian, it could subsequently recall complete and accurate images of them from partial and/or corrupted images, but it could not extrapolate any similarities between poodles and Pomeranians. Argentinian author Jorge Luis Borges's (1899–1989) famous short-story character, Funes "the Memorious," has a similar problem: Because of his prodigious memory, he literally cannot see the forest for the leaves. At one point the narrator explains, "He was, let us not forget, almost incapable of ideas of a general, Platonic sort. Not only was it difficult for [Funes] to comprehend that the generic symbol *dog* embraces so many unlike individuals of diverse size and form; it bothered him that the dog at three fourteen (seen from the side) should have the same name as the dog at three fifteen (seen from the front)" (Borges, 1942/1962, p. 65, italics in original).

In other words, neither a Hopfield network nor Funes possesses the ability to generate purposeful categories by which to classify input patterns, be they strings of ones and zeros or sensory percepts. In fact, the processing done by the Hopfield network is precisely the opposite of generalization: Its goal is to store specifics, not discern which details are salient and which should be discarded. Whether it represents correct recall or not, a Hopfield network will always produce a very specific memory pattern, like "the dog at three fifteen." This requirement of specificity severely limits this network's flexibility as a cognitive system: Because it cannot generalize, a Hopfield network cannot extract the rules that relate one item to another or assign different items to the same category. A Hopfield network essentially amounts to a pure, unstructured "storehouse" model of memory that does not account for the organized and structured aspects of memory.

The McClelland model overcomes this limitation of the Hopfield network with a second network component, a multilayer perceptron trained with backpropagation (see also chapter 2). Unlike the Hopfield network, the multilayer perceptron works by learning pairs of items, such as "poodle–dog." Prompted with the first, it will produce the second. If it is then presented with "Pomeranian–dog," it will not only learn this new pair but also group "Pomeranian" and "poodle" together as items that go with "dog." Its strength is precisely the Hopfield network's weakness: instead of specific instances, the multilayer perceptron learns categories. The perceptron is not constrained by the specifics of each pattern but can learn which parts of the patterns are important for classification and which other parts can be ignored. With every new input–output pair, the perceptron gradually creates a generalized, organized knowledge construct that can then be used for memory recall that is both flexible and efficient. It is the anti-Funes, a processor for which each pairing holds relatively little significance but which can learn to recognize easily that the 3:15 dog is the same as the 3:14 dog.

The major functional drawback of the multilayer perceptron trained with backpropagation is its slowness of learning. Repeated trials are needed for the generalizer to develop its categorizations, and it often requires thousands of presentations of a set of item pairs before it can accurately classify input patterns. "Dog" has to be an expansive enough category to accept breeds as different as "Doberman" and "poodle," but it also has to be discerning enough to classify "Persians" not with the "Pomeranians" and other dogs but under the heading "cats." Proper classification requires a large pool of examples and, as with the attainment of skill in real brains, it also requires lots of practice.

By combining these two systems, McClelland and colleagues obtain the best of both worlds. The Hopfield network "stores" specific instances and rehearses them for the perceptron, which eventually develops more generalized representations. The Hopfield network's attention to detail and verbatim reverberation make it the perfect teacher for the perceptron, as the former continually provides paired examples to the latter. The perceptron, meanwhile, converts specific examples from the Hopfield network into something more practical: generalized categories.

McClelland and colleagues' analysis is a *tour de force*. It applies the theory of reorganization derived from the multiple-memory systems hypothesis, explains the data from the past century of consolidation research, and creates a two-compartment neural network model that can be used to make predictions (see chapter 2). Their two-box model captures Squire and Cohen's view (1979) that memory consolidation can best be thought of as a reorganization of memory representations from one brain structure to another: Anatomically speaking, McClelland locates the buffer box in the hippocampus and the generalizer box in the neocortex, mirroring Squire and Cohen's theory. Also, the two-box model provides a more in-depth explanation of memory consolidation than previous models (McClelland et al., 1995).

The chief drawback of the McClelland model from a consolidation standpoint is that although it learns discrete input patterns, they are nevertheless treated in an anonymous way and aggregated together without regard for their potential relationships with one another. Although the theory of consolidation has evolved, the measure of the success of the consolidation process is still "how much memory?" as if memory were a bulk substance rather than composed of uniquely identifiable, manipulable, and potentially meaningful memory "items." Thus, a good metaphor for McClelland's model is still our pair of leaky buckets (see figure 2.1). Memory items pour into the first bucket—the temporary storage buffer unit—from which they trickle into the second bucket—the generalizer. The time it takes these liquid memories to move from the "buffer box" to the "generalizer box" is that of the consolidation process. Leakage is equivalent to forgetting. Both buckets leak, but the generalizer leaks less, mimicking the more durable condition of consolidated memories.

For memory scientists, this view is appealingly mechanistic; it fits well with what the empiricists Müller and Pilzecker (1900) were seeking, and to a certain degree it *is* representative of human memory characteristics. But this classic view is severely limited. A labile bucket and a stable bucket, no matter how extensively elaborated with neural network and adaptive learning capabilities, cannot explain why certain memories are or are not retained (and with what inaccuracies), or why the time course of consolidation is so variable, if it does not take into account the actual items being consolidated. It also offers nothing in the way of explanation of the effects of context, which scholars of collective memory value (see chapter 3).

In essence, the discrepancies between McClelland's model and noncomputational understandings of memory arise from the inability of the two-box or two-bucket model to address the *content* of memories. If memories are an amorphous mass of liquid in a bucket, there is no way to sort them or for them to exert subjective or context-dependent effects on each other. Put another way, the Bartlettian accuracy-based (how well) approach is incompatible with the classical, quantitative (how much) Ebbinghausian approach (see chapter 2). The data on retrograde amnesia, consolidation, and so forth, were gathered using Ebbinghausian techniques, and Müller and Pilzecker (among others) used those same techniques to quantify the trajectories of consolidation, retrograde amnesia, and other memory phenomena. Unsurprisingly, the data and the resulting model mirror the paradigm under which they were gathered and constructed.

Bartlett's qualitative approach, by contrast, asks not how much is recalled but how well the recollection matches the originally presented material. Most importantly, Bartlett's approach explores the specific ways in which the recollection is a distorted version of the original material (cf. Yonelinas, 2002). This framework, more subjective and contextual, attains greater explanatory reach for actually observed human memory foibles than quantitative methods, although it is admittedly more difficult and complicated to work within. The need to account for more qualitative phenomena, coupled with our review of humanities literature on collective memory, leads us to propose a more nuanced model, one that involves an additional functional element of relationality or meaning-making, as well as a contextual "entity" concept. Adding more complexity to the model helps to explain the more phenomenological, or subjective, findings, which cannot be ignored in any model that claims to describe memory consolidation comprehensively. Because the two-component model is clearly inadequate, we have added a third component to account for selection and relation in memory and a fourth, larger, all-encompassing component to contain the whole remembering entity—thus, three boxes in one.

Modeling Memory: The "Three-in-One" Model

We are finally ready to propose our own memory consolidation model. Although McClelland and colleagues' (1995) buffer and generalization modules can handle both detailed information and generalities deduced from it, the McClelland model lacks an intrinsic component responsible for establishing relationships between items. After all, the researcher sets the input–output pairs on which the McClelland model is trained, and in so doing he takes the place of an element that is needed to answer the essential questions that arise in consolidation. The buffer can temporarily hold an image, and the generalizer can learn to recognize it as a dog, but what establishes the relationship between the image and the category that the generalizer needs to learn? What establishes the relationships between categories that constitute useful knowledge? How is a dog related to other animals and objects? Of what actions is a dog capable? To answer such questions, an additional component is required, one that connects identifiable items to other items to create complex webs of associations. By creating connections between memory items, the relationality component of our model imbues memory items with meanings they could not possibly have on their own, as meaning is a relational concept (Hanks, 1996; Alberti, 2005).

The relationality component (relater) is an association-processing machine. This view is consistent with that of cognitive neuroscientist Howard Eichenbaum. Eichenbaum's work, which we will describe in detail in chapter 6, suggests that the main function of the hippocampus, and its main contribution to memory formation, is to identify, represent, and process associations between memory items. As Eichenbaum (2006) points out, the simplest relations are temporal: A leads to B leads to C. Webs of associations can be built up from simpler relationships: 1 leads to B leads to 3 is connected at B to the first string, so that 1 and 3 are related, through B, to A and C. The hippocampus is thought to represent such webs and access the associations they contain. In our model, the relater associates items temporarily stored in the buffer and presents them in properly associated, relationship-organized pairs to the generalizer, which extracts the rules by which the system of relationships can be efficiently represented.

Eichenbaum (personal communication) suggests that a useful metaphor for the relationality component is the search engine Google. Google's algorithm "discovers" how items are related by mining huge amounts of data gathered from the World Wide Web (Battelle, 2005). Using these data, Google forms an interconnected web of relationships that provides useful information about the world and is therefore meaningful. As people continue to create and link Web pages, they establish relationships among increasing numbers of Web sites, which Google incorporates into its ever-expanding web of associations.

Although Google, in some sense, is extremely efficient, its job is to create a raw web of associations, not to categorize or classify or otherwise generalize over items. One could imagine feeding pairs of Google-discovered associations to a giant multilayer perceptron, which would then generalize, in principle, over the entire contents of the World Wide Web. In our model, we imagine that the relationality component makes a smaller web of associations and feeds pairs of related items to the generalizer, which uses them to organize and structure stable memory.

In the same way that Google is valuable in itself, the relationality element is a versatile cognitive component that serves at least two additional functions over and above its role as trainer of the generalizer. First, the web of associations the relater discovers and maintains enables the entity to "surf" between items, producing a more complete picture of the world and the relationships between the items that compose it. Second, the links allow memory items to be recalled by context, and vice versa. American psychologist William James (1842–1910) referred to these relationships as "hooks" on which memories are hung, each connection being used to "fish [a memory] up when sunk below the surface. Together [these associations] form a network of attachments by which [a memory] is woven into the entire tissue of our thought" (James, 1899/1983, p. 77; cf. James, 1890, p. 662).

Unlike Google, the relationality element in our model, and presumably also in the actual consolidation processes of individual and collective memory, does not attempt to relate every available item. Instead, a critical function of the relationality element is selection, by which the element selects from among all available items those that can and should be related with one another. The selection feature is so critical to the function of this element that we also call it the selector/relater. As we will explain in the next chapter, the selector/relater often selects those items to which the entity pays particular attention, and that, in turn, is often determined by prior knowledge stored in the generalizer. Sometimes a particular item can be selected for verbatim storage in long-term memory. This process of modulation ensures that the item does not lose its specific characteristics as it is consolidated, while its relationships with categories in generalized memory are also represented.

These three elements—the buffer, the selector/relater, and the generalizer—together create a more complete model of memory consolidation. The first "box" stores items for the short term; the second selects items and establishes relationships between them; and the third extracts general themes that it stores for the long term. In our study, we further endeavor to make a distinction between the simple processes that happen automatically to memory items (here referred to as buffered storage, selection/relation, and generalization) and more complex processing. The simple processes could take place in a naïve entity; they are the irreducible elements required for consolidation. However, looking at only a single pass of the model from buffer to generalizer, from labile to stable, misses out on perhaps its most intriguing feature: recursion. Because the interactions between the elements of the model are two way, established knowledge can recur to influence the consolidation of new material.

The two-way interactions in the model account for the fact that consolidation is a dynamic process that integrates unconsolidated with already consolidated information. The basis for recursion is the already consolidated memories that can be variously described as knowledge, schemata, paradigms, frameworks, or narratives. These are all stable memory constructs whose rules are "in" the generalizer element. They can influence the consolidation process recursively by influencing which items pass from the buffer to the relater and how those items are associated by the relater. The looping, recursive nature of our consolidation model accounts for the fact that what is already remembered will influence the formation of new memories.

For our conceptual model to capture the full range of influences on the consolidation processes, it requires one more building block. The fourth "box" represents the entirety of the particular entity to which we are referring (figure 4.2). The "entity" provides context for the consolidation process. In individual memory studies, the entity is an individual person; for collective memory research, the entity is a particular social group. By an individual we mean the physical body and brain in which the consolidation process takes place. Scientists build computational models to exhibit the features of individual memory in human beings and other animals. These models, for simplicity's sake, tend to ignore many factors internal to the entity and all factors external to it. Such reductionism is a hallmark of modern science and has obviously proved useful for solving many problems. Notwithstanding, we want our model to have broader explanatory power and to account for the more subjective aspects of memory, so we cannot ignore the non-mnemonic influences on the memory consolidation process.

The entity box (i.e., the big box or the "one") represents the non-mnemonic factors such as the goals, plans, emotions, and desires of the individual or collective that can influence the memory consolidation process. The entity box also circumscribes these factors because,

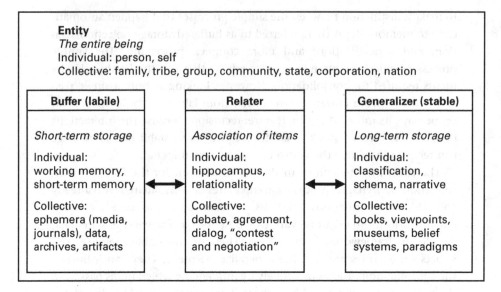

Figure 4.2
Our three-in-one model of consolidation. The three core elements of consolidation are the buffer, the relater, and the generalizer, which are embedded in the all encompassing entity element. The function of each element is underlined, and listed within each element's box are examples of structures or processes associated with that element.

for simplicity, it will not represent every such influence on individuals or collectives. Although it is true that individuals almost always exist in social conditions, we are trying to analogize *between* individual and collective memory, so we have lumped social influences on individuals with other external "environmental" factors. Thus on the one hand, when we apply the three-in-one model to individual memory formation, the "entity" box corresponds to a generic, biological individual, whereas social influences fall outside the big box. Again, we recognize this as a simplification and that individuals are always defined and shaped by social influences, but for the sake of our particular argument, we have bracketed those factors.

When discussing collective memory, on the other hand, the entities we will encounter include collectives such as religious and ethnic groups, socioeconomic classes, and nations. These are groups of individuals who share social frameworks. Families, organizations, laboratories, corporations, and professions including academic ones such as "scientists," "anthropologists," and "historians" are also entities with their own collective memories, social conventions, and characteristics. When humanists strive for the "context" of collective remembering, they are usually describing the entity that is the subject of their study. Collective entities have "outsides," too: The built and natural environments are external to these real and "imagined" communities (Anderson, 1983; Kansteiner, 2007). Also, like "society" for individuals, other collectives are a crucial part of the milieu of any particular collective. As scholars such as Michael Rothberg (2009) have shown, other collectives' memories of events experienced by a particular group can be part of that group's context. Thus, French memories of colonialism interact with French-Jewish memories of the Holocaust, which interact with the many different American memories of slavery. We acknowledge the outside influences on collectives, but when we apply the three-in-one model to collective memory formation, the "entity" corresponds to that collective, whereas outside influences fall outside the big box.

These simplifications would not prevent application of the three-in-one model to interactions between individuals, between collectives, or between individuals and collectives. They would simply be seen as occurring on the level of their respective entities. For example, two entities could be in competition, and this interentity interaction could affect the goals, plans, emotions, and desires of each entity. Those non-mnemonic factors would, in turn, affect memory consolidation in each entity, just as they are intended to do in the three-in-one model. Assuming that outside influences fall outside the big box is equivalent to assuming that they do not affect memory consolidation directly, but only indirectly through interactions occurring at the entity level. This simplification is not prohibitive and may actually correspond with the way outside influences affect real memory consolidation in individuals and collectives.

The specter of a homunculus, or "central executive" (Baddeley & Hitch, 1974), "pulling the strings" of the mind has bedeviled psychologists for decades. It may seem that bringing the entity into the consolidation equation is a backhanded way of succumbing to the desire for an inscrutable consciousness, but that is not the case. Positing a contextual entity merely allows us to account for non-mnemonic influences. Meanwhile, the entity is not the only component that can influence the consolidation process. The recursion of stable knowledge from the generalizer automatically contributes to the organization of the labile items that are then consolidated, so the generalizer influences its own development, just as collectives' narratives, though changing, encourage consistency and coherency in their content (H. V. White, 1987; Kearney, 1997). Therefore, in our proposed model, already consolidated memories influence what is stored in the first element and drive selection and relation of items in the second element, thereby influencing the content of the generalizations generated in the third element. The three consolidation elements are "awash" in the consolidating entity, which can influence the consolidation process according to its goals, desires, and emotions. The interactions between the buffer, relater, generalizer, and entity are complicated, but the pay-off for examining them is greater insight into the formation of individual and collective memory. We examine them in detail in part II. Before we get there, we provide a quick foretaste of how our model can be applied to collectives.

Consolidation of Scientific Memory: Take 1

We can illustrate the application of our model on a collective level with a reinterpretation of a familiar theory, that of scientific paradigms and revolutions, as an example of the consolidation of what can be considered "scientific collective memory." Thomas Kuhn (1962/1996) defined a "scientific paradigm" as a framework that contains facts, concepts, laws, theories, and points of view that cannot be divided into its separate components. As an "implicit body of intertwined theoretical and methodological belief" (pp. 16–17), a paradigm, such as heliocentricism or quantum mechanics, serves as the basis for decision and action for a collective of scientists. Frederic Bartlett and Maurice Halbwachs would have recognized scientific paradigms, respectively, as types of schemata or social frameworks.

Their paradigm largely determines for scientists which experiments they do and which types of facts they gather under its rubric of "normal science." It influences the design of their experiments and apparatuses, the character of their observations, and the particularities of their theories. Moreover, scientific paradigms are taught holistically, in that their facts, theories, and applications are codified together in textbooks and conveyed more or less unquestioned to students. Thus, a scientific paradigm can be thought of as a collective memory whose primary stable knowledge (memory) structures are science textbooks. As with all memory structures, it is not their physical characteristics but their place in the organization of memory that determines whether they serve labile or stable memory (see chapter 1). That digital versions of textbooks are becoming increasingly available does not change their importance as consolidated structures supporting stable, collective scientific memory.

Not only is a paradigm an example of collective memory but also the process of formation of a scientific paradigm demonstrates collective consolidation. James Wertsch (2002) has observed that collectives develop usable knowledge constructs through a process that involves "contest and negotiation." Kuhn similarly describes how a scientific paradigm is formed through a process that laypeople might suppose to be uncharacteristic of scientists: "History suggests that the road to a firm research consensus is extraordinarily arduous" (Kuhn, 1962/1996, p. 15). Kuhn's ideas on the development of scientific thought can be recast in terms of our memory consolidation model: Data and other facts, held in a labile state, are arranged into a consistent set of relations that is recorded in a more general and stable form, and the whole process is directed both by the recurrent influence of the already consolidated paradigm and by the goals and desires of a particular scientific community acting as an entity.

More specifically, data and reports on them constitute the labile memory of a scientific community, whereas textbooks (and related materials such as course curricula) constitute stable memory (knowledge) structures. The scientific community itself constitutes the relationality element, as individuals compete with each other, through written communications and at conferences, to find the most elegant and powerful explanation for the facts at hand. The tendency is to find explanations that are consistent with the reigning paradigm, but that is not always possible, as we consider further in the next paragraph. That the direction of research may be driven in part by more prosaic concerns, such as the availability of research funding, can be chalked up to the influence of the consolidating entity.

Kuhn, of course, was concerned less with the formation of an initial paradigm than with the re-formation of an existing paradigm through a "scientific revolution," which he conceived by analogy with political revolutions. The greatest threat to an existing paradigm, Kuhn suggests, is the accumulation of new data that do not fit it. Because a paradigm can answer most, but not all, of the questions it proposes, the types of anomalies that cause paradigms to change are "the recognized anomalies whose characteristic feature is their stubborn refusal to be assimilated to existing paradigms" (Kuhn, 1962/1996, p. 97). At first, practicing scientists who were trained under, and are invested in, the reigning paradigm disregard the stubborn facts in favor of the compliant ones. Put another way, the group selects and consolidates items that can be consistently related to one another under its paradigm and "forgets" the "noncompliant" rest. However, inconsistent facts can only be overlooked for so long, and eventually someone—frequently a young researcher or else a newcomer to the field who is not so invested in the current paradigm—reinterprets the "anomalies" as the basis for a new paradigm that engenders a completely different way of looking at the body of data.

The consolidation of a new framework, this "paradigm shift," turns on relationality. Kuhn quotes British historian Herbert Butterfield (1900–1979), who described paradigm change as "handling the same bundle of data as before, but placing them in a new system of relations with one another by giving them a different framework" (as quoted on p. 85 of Kuhn, 1962/1996). The consolidation of collective memory always builds on the prior establishment of a system of relations, but the creation of a new paradigm requires "a new system of relations," a new way of explaining observations in terms of each other.

Kuhn notes that it often takes a generation for a new paradigm to become dominant, as adherents to the old paradigm convert or die off, while the youngbloods grow up with it. For this reason, perhaps the most important outcome of a scientific revolution is the writing of new textbooks, because these provide an introduction to future scientists of the history of their re-imagined field and the possibilities for normal research within it. Because they are involved in the intergenerational transfer of knowledge, textbooks reflect the stable collective memory of a particular scientific community. As we will explain in detail in chapter 7, temporary storage (buffering), relation, generalization, and other features of memory consolidation are involved in textbook writing, and the reigning paradigm, itself a form of generalized memory, can recur to influence all of these processes.

Thus, paradigms are the social frameworks that shape stable, if not permanent, scientific collective memories. Some historians of science have criticized the idea of abrupt scientific revolution and have noted that scientific understanding usually changes more gradually (M. J. Klein, 1972; Krige, 1980; Lindberg & Westman, 1990; Pickering, 1995; Hoddeson, 2007). Yet many of Kuhn's observations on the progress of scientific thinking are nevertheless relevant to a theorization of the consolidation of collective memory. Criticism of Kuhn's theory on the basis that real science changes gradually, rather than through revolution, does not diminish its usefulness to us as a description of collective memory consolidation, which is faster or slower, depending on many factors, as McGaugh (1972) suggested for individual memory consolidation. Knowledge structures such as textbooks still support both the normal science of researchers and the transmission of the paradigm to the next generation of scientists. A paradigm shift at whatever speed requires the selection of what were previously thought to be "anomalous" observations and their relation in a new way. This relational process changes the paradigms, either abruptly or gradually, resulting in further generalization over facts that improves their explanatory power. The process of consolidation of scientific collective memory is thus part of a larger process that dynamically structures scientific knowledge.

As we will show in part II, our three-in-one framework can be applied to memory consolidation in individuals and in collectives of many different kinds. The underlying theme is that the process of memory consolidation, involving buffering, relation, and generalization, under the influence of an encompassing entity, is essentially the same on these different levels. Our main goal in part II is to describe structures and processes that correspond to the elements and functions represented in our three-in-one model on individual and collective levels. The structures, of course, are different, but the processes are analogous.

Conclusion

The first three chapters provided a brief overview of individual memory consolidation and of collective memory as a phenomenon in itself. This chapter provided an outline of our model of memory consolidation in individuals and collectives. Along the way we encountered but a few of the individuals who have laid the groundwork for our model. After Hermann Ebbinghaus began describing the quantitative aspects of memory, Frederic Bartlett explored its more qualitative aspects. Georg Müller and Alfons Pilzecker discovered that memories are not formed immediately but are transformed from a short-term to a long-term state through a process that they termed "consolidation." The concept of consolidation was applied to explain the phenomenon of retrograde amnesia, which William Scoville and Brenda Milner observed in the patient H.M. after Scoville had surgically removed most of H.M.'s medial temporal lobes (including his hippocampi) in order to treat his epilepsy. Although H.M. could still remember events from his childhood, his memory for more recent events, which were still in the process of consolidating, vanished. This led to studies by Larry Squire, Neal Cohen, Howard Eichenbaum, and many others indicating that the hippocampus is a brain structure of central importance for the process of consolidation.

Despite extensive research over the past 100-plus years, memory remains a wonderfully mysterious phenomenon. Consequently, researchers have used, and continue to use, metaphors to help them understand it. Of all the metaphors conceived for memory formation, the most powerful have been neural network models, and we describe some of them at a level of detail appropriate to their importance. Principle among these is the two-box (two-network) model of James McClelland, Bruce McNaughton, and Randall O'Reilly. In this model, short-term memories are held in a dynamic network, which can repeatedly present verbatim memory items to a static network, which can learn to produce a generalized and efficient representation of the items that constitutes long-term memories. The model of McClelland and colleagues was able to explain many phenomena associated with memory consolidation in individuals.

Scholarly work on collective memory begins with Maurice Halbwachs, who argued that individual memory is mediated by a form of collective memory he called "social frameworks." Whereas some contemporary scholars, such as Jeffrey Olick, recognize a "strong" form of collective memory, which is more than the sum of the separate memories of the individuals that compose the collective, others, such as James Wertsch, consider only a "weak" form of collective memory, which is more or less the same version of the past that most members of a group share. According to Wertsch, collective memory is constrained to be weak because it is necessarily distributed over the members of the collective, but we argue that synergistic, and therefore "strong," forms of memory can emerge in systems (such as neural networks) composed of interconnected and "distributed" elements. In either case, whether collective memory is more than the sum or more or less shared, it must somehow be formed. We have expanded the two-box model to a four-box (or, more precisely, three-in-one) model to account for the formation, or consolidation, of individual as well as collective memory.

The two-box model of McClelland and colleagues consists of a "buffer," which temporarily holds memory items over the short term, and a "generalizer," which holds them in a compressed and efficient manner for the long term. The generalizer does not merely cluster similar items together. Even in its simplest form, which is a classifier, the generalizer stores not only general categories but also the relationships between specific items and the classes to which they belong. In order for the generalizer to learn these relationships, it must be presented with items organized according to them (as in member–class pairs such as "poodle–dog"). To account for this function we add, in between the buffer and generalizer, a "relater" element that establishes relationships between items and presents items in a relationship-organized fashion to the generalizer. The relater in our model corresponds to the hippocampus, which is a neural structure that is central to the consolidation process and which, as Howard Eichenbaum has shown, identifies and manages associations between memory items in real brains.

As Eichenbaum has suggested, an appropriate metaphor for the hippocampus is Google, which accurately determines the relationships that associate Web sites according to the actual arrangement of links between them. However, as Frederic Bartlett has shown, individual memory is fraught with inaccuracies, and individuals idiosyncratically distort the relationships they remember. This leads us to add a fourth, overarching component, the "entity," which accounts for non-mnemonic factors such as goals, plans, desires, and emotions that can influence the core elements of the memory consolidation process, especially the relater.

Our three-in-one model is therefore composed of a buffer, a relater, and a generalizer, all contained within an encompassing entity, and we use this model to help us establish an analogy between individual and collective memory consolidation. In part II of this monograph we provide evidence that memory consolidation involves essentially the same processes on both individual and collective levels. Specifically, whether in individuals or collectives, memory items are stored for the short term in the buffer, relationships among them are identified by the relater, which repeatedly presents them, in a relationship-organized manner, to the generalizer, which stores them in a compressed and efficient form for the long term. All stages of the consolidation process, especially the relater stage, can be influenced by the goals, plans, desires, and emotions of the consolidating entity.

Historian Alan Confino (1997), among others, has issued a charge to collective memory researchers to develop a coherent theory of memory that can move the field beyond yet another description of the construction of (collective) memory to something approaching a functional, explanatory model. By offering an explanation for memory formation on both individual and collective levels, we believe we have developed that model.

II The Memory Consolidation Process

5   Buffering and Attention

This chapter explores the concept of "temporary storage" in memory while building a case for the buffer element in our model of memory consolidation. As we explained in our model overview in chapter 4, memory consolidation involves the creation of less-changeable (stable) generalized knowledge constructs from systems of relationships that are established between items held in a more-changeable (labile) fashion. Because of the recursion of already consolidated memory, stable items are also worked into these systems of relationships. To arrange the items for consolidation, they must be held in such a way that they can be easily manipulated. The main purpose of the buffer in memory consolidation, therefore, is to hold various memory items to allow for further processing (specifically selection, relation, and generalization). Metaphors for the memory buffer include data buffers in computers, scratchpads, tabletops, workshops, or any space in which items can be provisionally laid out and easily accessed as they are assembled into more permanent structures. In this chapter, we will discuss the process of buffering in individual and collective memory formation.

Although the buffer can hold long-term memory items that are downloaded to it from the generalizer, its main function is to hold short-term memory items. Two specific types of short-term memory have been identified in individual brains: immediate memory and working memory (Eichenbaum, 2003a). Immediate memory holds memory items over a span of seconds to minutes. Working memory is that portion of immediate memory in which temporarily stored items are actively processed by other cognitive subsystems (such as reasoning). For example, you use immediate memory when you remember a new phone number long enough to dial it, and you use working memory when you represent the various terms as you mentally work out the value of x in the equation $2x + 3 = 7$. Working memory is a subset of immediate memory, which is a subset of unconsolidated memory. Nonimmediate, unconsolidated memory is everything in the buffer that is not immediate (or working) memory. For example, you use nonimmediate, unconsolidated memory when you remember what you received for your birthday last month. Over the long term (i.e., years), the details will probably be forgotten, but for now they have left a recoverable trace that is not held in immediate (or working) memory.

The neural structures that support the buffer in individual brains are located in the cerebrum (Eichenbaum, 2003a). Immediate memory and working memory are mediated mainly by the cerebral cortex, especially the prefrontal cortex. Nonimmediate, unconsolidated memory appears to be mediated by the hippocampus and related medial temporal lobe (MTL) structures (probably through their interactions with the cerebral cortex) because this form of unconsolidated memory, but not immediate (or working) memory, is lost when the hippocampus and related MTL structures are removed (see chapter 2).

Another important function of the buffer is to provide an arena for attention. Attention is a significant factor in consolidation because it increases the probability that short-term items will become consolidated into long-term items (Milner, 1999). Neuroscientifically, "attention" has not been precisely pinned down, but current research suggests attention in individual brains is mediated mainly by the prefrontal cortex through its interactions with various other cortical and subcortical structures (Knudsen, 2007). We use the term "attention" to connote the process by which an entity focuses on certain items, and not on others, in the memory buffer. The buffer can temporarily hold memory items both to which the entity attends and to which it is not currently attending. For example, you are able to remember what you ate for breakfast this morning and where you parked your car, now that we are bringing these items to your attention, even though you have not been attending to them constantly since you began your day. Like all labile memories, those concerning aspects of your morning activities are stored in your buffer whether or not you attend to them, but they are more likely to be consolidated, into stable memories, if you do attend to them.

In our model, we assume that the entity focuses attention onto specific items in the buffer and that its attention is directed to those items according to its non-mnemonic properties (e.g., goals, plans, desires, emotions, etc.) or through activation of already consolidated memory in the generalizer. Once directed, the entity can "tag" certain items in the buffer for attentional focus, and the buffer can register this tag. It is important to reiterate that buffers do not only hold unconsolidated and unfamiliar items; already consolidated items, from the generalizer, may also be temporarily represented in the buffer. The generalizer therefore can act recursively on the buffer both by sending already consolidated items back to it and by directing the entity to focus attention on specific items in the buffer. Because some items can be consolidated before others, it is not always the case that unconsolidated items are "new" and already consolidated items are "old," but for ease of expression in discussing subtle phenomena such as attention, we will sometimes use "new" and "old" to refer to unconsolidated and already consolidated items, respectively. Thus, attention increases the probability that new items will be consolidated and that old items' traces and connections will be strengthened.

The first half of this chapter presents a semihistorical outline of the functional aspects of the buffer in individual brains. We will examine the phenomenon of "chunking," the "magical number seven," and working memory mechanisms such as the "phonological loop" and "visuospatial sketchpad." Because we are particularly concerned with memory consolidation, we will emphasize the effect of existing, consolidated knowledge on the buffer. This recursion explains not only the representation of old as well as new information in the buffer but also how the generalizer can "automatically" direct the entity's attention without resorting to a Maxwell's demonesque homunculus or "central executive."

Following our analogical reasoning, there must also be buffered storage in collective memory for those collective experiences and ideas that are labile but still neither consolidated nor forgotten. The well-established theory in the humanities of the "public sphere" most closely captures the function of a collective buffer for labile memories and for contest and negotiation over their meanings. Mass media are major players in collective buffering today, providing substrates for consolidation, holding the "ephemera," directing the public's attention, and responding to its desires. The second half of this chapter therefore uses three case studies that reveal the nature of the collective memory buffer. We will delve into archives and archival practices, "the news," and the public-art movement known as "countermemorialization." As we will see, all these social and cultural institutions involve the same sort of recursive and attentional effects that have been described for the individual memory buffer.

Hold That Thought! Temporary Storage in Individual Memory

All individual memories, whether labile or stable, must have some neurobiological underpinning. Most neuroscientists now agree that long-term (stable) memory involves long-lasting changes in the strengths of the synapses between neurons (e.g., Bailey & Kandel, 1993). The neurobiological basis of short-term memory is more of a mystery, which is captured by a quotation from neuroscientist Endel Tulving. As he put it in an interview:

As a scientist I am compelled to the conclusion—not postulation, not assumption, but conclusion—that there must exist certain physical-chemical changes in the nervous tissue that correspond to the storage of information, or to the engram, changes that constitute the necessary conditions of remembering. (The alternative stance, that it may be possible for any behavior or any thought to occur independently of physical changes in the nervous system, as all your good readers know, is sheer mysticism.) However, if the engram is a kind of entity that manifests itself only in activity, or retrieval, then we might conjecture that the physical changes resulting from experience do not exist as an engram in the absence of that activity. And we can also imagine that the engram, qua engram, is not detectable in its quiescent state, that is, in the absence of retrieval, with any physical technique. (attributed to Tulving in Gazzaniga, 1997, pp. 98–99)

We already encountered (see chapter 2) the postulate by neuropsychologist Donald Hebb (1949) that short-term, labile memories could be stored in networks of firing neurons that connect to each other in loops, forming "reverberating circuits." Once this circular activity stops, the memory trace (i.e., engram) is lost. Hebb envisioned that these reverberations would eventually "potentiate," or strengthen, the synaptic connections between the firing neurons, leaving a lasting synaptic engram (cf. Lashley, 1950; for a review, see Schachter & Scarry, 2001). It is possible that reverberation corresponds to short-term memory, whereas potentiation corresponds to long-term memory, but it is also possible that both neural activity and synaptic weight changes are involved in short-term memory. In support of this view are findings that certain forms of short-term memory are associated with temporary increases in the activity of neurons in the frontal, parietal, and temporal cortices, which can be simulated using neural network models composed of units arranged in circuits that can "learn" to activate one another by changing their connection (synaptic) weights (Fuster et al., 1985; Gnadt & Andersen, 1988; Zipser, 1991). In any case, we are less interested in the biology than in the phenomenon of short-term buffering in individual brains because we wish to make a functional rather than a structural analogy between individual and collective memory. The study of the functional properties of short-term memory in individual brains began with the discovery of its limitations.

Experimental psychologist J.R.M. Hayes, with his investigation of the capacity of immediate memory, laid the foundation for the study of temporary storage in the early 1950s (Hayes, 1952). Hayes found that individuals were able to hold approximately seven items in immediate memory with great accuracy, whereas higher numbers of items presented more of a challenge. This set the stage for American psychologist George Miller's 1956 paper, "The Magical Number Seven, Plus or Minus Two: Some Limits on Our Capacity for Processing Information." Hayes had shown that people could hold seven tones, seven words, or seven digits in their immediate memory. The extension Miller made was that the complexity of the items was essentially *unrelated* to the number of items that could be recalled. He noticed that subjects could remember seven letters *or* seven words, even though remembering seven words obviously required remembering more than seven letters. Similarly, subjects who studied a rule for coding binary sequences as digits were able to increase their memory for binary sequences manyfold (Miller, 1956). Miller called this ability to extend the capacity of immediate memory "recoding." He postulated that low-level items (e.g., letters) are grouped into high-level "chunks" (e.g., words), and that individuals can retain seven chunks in immediate memory.

The "chunking" breakthrough confirmed that the capacity of immediate memory was limited, but only in terms of number of items that could be retained in it. The complexity of the items was limited only by the amount of a priori "recoding" that could be applied to the items. By analogy, you could hold an onion in one hand and a chicken leg in your other hand or, because of "recoding" in the form of food processing, you could hold in one hand a can of chicken soup and in the other a can of beef stew. Although the "capacity" of your hands is two in both cases, you hold many more ingredients in the latter case. Similarly, the capacity of immediate memory is limited to a set number of items, but the processing power of other cognitive subsystems can extend the complexity of those items. In our model, the generalizer provides most of this additional processing power. This is just one example of the influence of other components on the functioning of the buffer.

More recent research suggests that Miller and his colleagues may have set the number of memory chunks too high. Modern estimates of the maximum number of concurrent chunks place the number at about four plus or minus one, though this depends on experimental protocol and the character of the chunks (Murray, 1968; Baddeley & Hitch, 1974; for discussion, see Cowan, 2001, 2005). Whatever the number of chunks, the key facet of recoding for our argument is the stabilizing force it exerts on memory items. Whereas a chain of letters like CBSMTVIRSSUVTNT might be difficult for labile memory to hang onto, once the letters are chunked into CBS-MTV-IRS-SUV-TNT, the task becomes much easier, and what never would have made it into stable memory has a greater chance of consolidating.

Discovery of the phenomenon of chunking raised new questions. How are chunks created? How do we choose what to focus on from the vast amount of input to immediate memory? To answer them, psychologists Alan Baddeley and Graham Hitch elaborated a multisystem model of immediate and working memory (Baddeley & Hitch, 1974; see also Schacter & Tulving, 1994). They developed a concept called the "central executive" to coordinate the activities of a set of short-term memory systems, each with its own specific functions. One of these systems, the "phonological loop," stores and maintains recent verbal information and makes it available for cognitive processing. For instance, subjects can hear words read aloud and, through the phonological loop, maintain the sounds they had heard long enough to provide definitions for the words. A similar system, the "visuospatial sketchpad," acts as a temporary representation area for visual data. This allows subjects to "picture in their mind's eye" an event or place described to them in words (either aloud or in text). For instance, a subject could imagine the layout of a room based on a written description and maintain that mental picture via the visuospatial sketchpad long enough to draw the scene. In that cognitive processes act on these representations, the phonological loop and the visuospatial sketchpad are intended as models not only of immediate memory but of working memory as well.

In his 2000 follow-up paper, Baddeley added yet another short-term memory system to the model: an "episodic buffer" that serves as a "space" in which to construct a sequence of remembered events from the phonological loop and the visuospatial sketchpad. For example, subjects could recall the sounds and sights of an episode and assemble them into a rough chronology using this buffer. All of these systems act as "slaves" that store and organize the items in them according to commands from the central executive. Although the fundamental purpose of this multisystem working memory arrangement is short-term storage, the central executive acts as a "supervisory attention system" (Norman & Shallice, 1986) to intervene in the other three systems' processing. It can change the cognitive functions that are applied to the items as well as allocate attention between the three other systems. Although the central executive remains hypothetical, clinical evidence supporting its existence comes from patients with a "disexecutive syndrome" resulting from damage to the frontal lobes. These patients show action-control deficits and impulsivity, which seem consistent with the loss of a master control, supervisory attention system (Shallice, 1988).

May We Have Your Attention, Please!

According to Baddeley and Hitch's formulation, the central executive is external to the storage systems and apparently uninfluenced by them. Having appealed to this "higher power" to explain attentional allocation in working memory, they have essentially black-boxed the central executive. This begs the question: What are *its* functional principles? Is it possible to conceive of an internally coherent model of attentional allocation in working memory that requires no "string puller"? Both in the case of the Baddeley and Hitch short-term memory model and in our memory consolidation model, we would suggest that it is in fact possible. In the case of our three-in-one model, where the entity is the "one," it should be possible for the entity's attention to be directed entirely by the other three, internal, lower-level, strictly mnemonic components, namely the buffer, the relater, and, especially, the generalizer.

Key to this internal regulation is the buffer's ability to handle labile as well as stable representations and the two-way interconnections between all of the elements in the three-in-one model. Miller's recoding theory already suggests that chunks, which increase the total information content of working memory, are an amalgam of both new and old memory items. In our model, when a new item enters the buffer, it would activate already consolidated items stored in the generalizer, which could download the activated, old items to the buffer. This interaction could occur directly between the buffer and the generalizer (a double-headed arrow between buffer and generalizer has been omitted from the model diagram in figure 4.2 for clarity). Now the new item and the old items are all held in the buffer (where they could be chunked or otherwise manipulated). It is easy to imagine new items activating old items, and that the old items so activated could automatically direct the attention of the entity without the need to postulate a hidden decision-maker—no central executive necessary.

This is precisely the critique that computational neuroscientist Randall O'Reilly levels against Baddeley and Hitch's mechanism of attentional allocation. In a paper entitled "Banishing the Homunculus: Making Working Memory Work" (Hazy et al., 2006), he and his colleagues posit a three-part model: a working-memory element that stores and maintains active representations of items; an adaptive, reinforcement-guided mechanism that learns to select between competing working memory representations (in effect, an attention allocator); and a long-term memory store that provides a source of established representations from which the working memory element draws. Performing in concert, the first two elements update and maintain labile representations that activate the associated representations in long-term memory. These stable, long-term representations in turn act upon the first two components to permit attentional allocation according to already acquired knowledge. Although the working memory element closely resembles the ones Baddeley and Hitch propose, its computational power is extended by introducing the prior knowledge provided by long-term memory and by the attentional allocation capability of the reinforcement learning system.

The resulting model is quite effective at simulating, and explaining, human performance on a variety of common laboratory tests—such as the Stroop task, the 1-2-AX task, and the Wisconsin card-sorting task—all of which test a subject's ability to attend correctly to specific items in working memory.[1] Although it may seem mechanistic to humanists, the model is worth mentioning because it resembles our memory model in four important respects. First, the working memory element of the O'Reilly group's model, like our buffer element, handles both new and old items. Second, specific items in working memory can be selected from among other items for specific purposes (see chapter 6). Third, selective attention is subject to the recursive influence of already consolidated memory. Fourth, both models are based on actual observations and therefore provide realistic views of the buffer element of individual memory.

Buffered Storage in Collective Memory

The first part of this chapter described the buffer in individual brains as an element of the memory consolidation system. All labile memory items can be held temporarily in it, but some can be singled out for special

---

1. The Stroop task (Stroop, 1935) involves differentiating between two kinds of information, one of which is relevant for the task (i.e., reading the word for a color) and the other of which is a confounding stimulus (i.e., letters in a different color). The 1-2-AX task (Frank et al., 2001) requires subjects to make decisions about a current stimulus based on a previous stimulus (i.e., if a 1 was presented, then they should respond only when the sequence AX comes up, but if a 2 was presented, then the target sequence is BY). The Wisconsin card-sorting task (Berg, 1948) requires subjects to discern an unspoken rule for correctly matching cards while being provided feedback on their performance; the rule changes periodically and the subject must discern the new rule.

attention. It also briefly explored the neural substrates of the neurobiological buffer and found that they may involve changes in both the structure (synapses) and dynamics (neural activity) of the brain. Analogous themes emerge in analyzing buffered storage and attention on collective levels. In the humanities, a tension exists between "material culture" and "discourse" in society, which would correspond roughly to structures and processes within the more limited context of collective memory. We touched on this issue already with our discussion of cultural tools (see chapter 3). The ways in which cultural tools such as books and newspapers, photographs, films, and monuments are treated as objects fall within the domain of material culture studies. In contrast, the ways in which groups in society deploy cultural tools in assertions and arguments about truth, identity, and other systems of meaning subject material objects to discursive analysis.

The material-culture/discursive dimension of society is independent from the temporal short-term/long-term dimension of collective memory. As a consequence of this independence, the same item of material culture can serve as a memory structure supporting either short-term or long-term collective memory, depending on whether the discourse in which it is deployed involves short-term or long-term collective remembering. Imagine, for example, a single artifact, discovered by a team of archeologists who at first are unable to place it within any context. Though it may be ancient, the artifact itself and the discourse surrounding it are at that moment part of the archeologists' short-term collective memory. This memory is short term because it is labile: It is meant to be a temporary interpretation that is subject to change. As the archeologists uncover more artifacts from the same site, relate them together, read the pertinent literature, and interact with colleagues in their field, they gradually are able to construct a narrative that places the artifact in context, and the artifact and the discourse surrounding it become associated with the long-term collective memory of archeologists in that field. This memory is long term because it is stable: It is meant to endure as a well-worked-out interpretation. Note that the artifact has not changed materially, but the discourse surrounding it transitions from short-term (expected to change) to long-term (intended to last) collective memory. As this example illustrates, it is not the physical characteristics of a memory structure but the way it fits into the organization of memory that determines whether it serves short-term or long-term remembering.

Historians are familiar with the idea that for a particular event—the English Peasants' Revolt of 1381, say—there are more and less durable forms of evidence. Whereas the physical characteristics of an object cannot determine whether it serves short-term or long-term memory, they obviously can determine whether or not historians of a later age have access to it. Less durable forms of evidence of the revolt included women's gossip, bards' tales, and elite observers' verbal reports (e.g., Dobson, 1970). Although the relative reliability of these sources may have been open to debate, the fate of the evidence they provided was not: Because the goodwives, the bards, the observers, and the revolting peasants all eventually died, the information they stored—in their neural tissue, and expressed in their songs and conversations—was lost.

The reason we know anything at all about the uprising today is due to the durability of parchment. The monk also died, of course, but Thomas Walsingham (1376–ca. 1422) (for that was his name) left one of the written records that provide the basis for our understanding of those events today. His report, in his time, was probably considered "elite"; by the 12th and 13th centuries in England, eliteness no longer depended on one's ability to remember oral tradition, but rather on one's ability to commit something to writing (Clanchy, 1979/1993). It was probably also considered reliable, but we can only speculate as to whether his report, in his time, served short-term or long-term remembering. Nowadays, Walsingham's text resides in the collective memory buffer, along with countless other objects, documents, tape recordings, films, digital media, and so forth, that have been preserved for posterity. There they serve as short-term memory items and are available for selection and relation by researchers and scholars who will consolidate generalized, long-term memories from them.

Archives as Collective Memory Buffers

What is left for historians to work with years or centuries later is conventionally known as an archives. Scholars distinguish between "an archives," which is a collection of items, and "the archive," which is the whole enterprise of archiving that includes the archivists and the institutions that support them (Nesmith, 2002; Stoler, 2002). The reason for making this distinction is to point out that an archives does not make itself: Any actively organized collection of items will be influenced by the goals, plans, and desires of the archivists who make it. Furthermore, the archivists themselves will be influenced by the institutions that support them and by the social frameworks established by the larger society of which they are a part.

French philosopher Jacques Derrida (1930–2004) famously begins his
discussion of "the archive" (Derrida, 1995/1996) with the two meanings
of the Greek word ἀρχεῖο: a "commencement" and a "commandment."
The archives is where knowledge begins; it is also where power exerts itself
on knowledge acquisition by dictating what is archived and what is not.
But acknowledging that archivists are influenced by social frameworks and
institutional biases does not mean that what they produce, an archives, is
a form of consolidated collective memory. On the contrary, despite
unavoidable bias, an archives is intended to serve as an unconsolidated
source of raw material for the research and scholarly communities to
process.

Archivist Tom Nesmith (2002) promotes a metaphor for the role of the
archivist that he attributes to Douglas Brymner, who was the first head of
what is now the National Archives of Canada. According to this metaphor,
an archivist is one who clears the land in preparation for another, such as
a historian, who cultivates it. In terms of our model, an archivist is one
who brings items together into the buffer in preparation for others, including historians, to select items from it, to relate them together, and to
condense those relationships into generalized knowledge. The idea that
the collective buffer could be influenced by social frameworks, which are
contained in the collective generalizer, is analogous to the influences of
long-term memory on short-term memory that we described for individuals in the first half of this chapter. That all of the processors, the buffer as
well as the selector/relater and generalizer, are awash in the influence of
the entity is consistent with our three-in-one view of individual and collective memory consolidation. Although an archives is intended as a form
of collective short-term memory, the influences on it from the collective
generalizer (e.g., social frameworks) and the collective entity (e.g., the
institution) nevertheless influence its contents and thereby influence the
long-term memory that is ultimately consolidated from it.

Although "archives" has traditionally referred to a collection of documents, other postmodern scholars have reinterpreted "archive" to mean
any actively collected repository of items (Foucault, 1969/1972, 1966/1971;
Burton, 2003, 2005; Halberstam, 2005). This redefinition broadens the
scope of what should be considered as suitable material for archiving. In
this way, for instance, the scrapbooks about state fairs, which their owners
allow a historian to use, constitute an "archive" just as much as does a
rigorously sorted and indexed collection of government documents.
Either way, the contents of such cultural tools must enter the collective
buffer and must furthermore become part of collective working memory

(the part of the buffer being accessed by other processors) in order to be consolidated.

Often, the interpretation of the contents of an archives depends on the era during which those contents were admitted to it. Historian of science David Kronick (1917–2006) recognized this in the context of his archival source of choice, early modern scientific periodicals. He noted that they "served two primary and important functions: first as a vehicle and then as a depository" (Kronick, 2004, pp. viii–ix). As "vehicles," periodicals such as the *Journal des Sçavans* (founded January 1665) or the *The Philosophical Transactions of the Royal Society* (founded March 1665) disseminated information through the physical and discursive sphere. Back then, they were objects that supported short-term collective memory for the early modern European science community, the material substrate for their discussions about magnifying lenses, brain anatomy, color theory, and gravity. In modern times as "depositories," the periodicals again serve short-term, scientific collective memory because members of new collectives access their content. Historians who use them now as sources for their research re-inject their content into the collective memory buffers of contemporary historians and scientists, who may consolidate new memories from them. So archives contribute to collective memory formation as resources through which items, events, and ideas are made available to the ongoing collective consolidation process.

The history of specific cultural tools is an important part of collective memory research. Why do we have access to certain sources (objects, memory structures) and not others? Whose voices are silent, or of what discourses are we ignorant, because of the ways in which some sources have become depositories but not others (Derrida, 1995/1996)? Over the course of the twentieth century, historians and cultural theorists increasingly complicated their understandings of the "biographies" of historical sources—that is, their trajectories from made objects or written documents to archived sources (Alberti, 2005; Kohlstedt, 2005). For instance, French historian Marc Bloch (1886–1944) wrote in his extended essay, *The Historian's Craft* (1953/1964; published posthumously), of the factors that affect which private sources become public—namely, contingency, catastrophe, and "the spirit of the secret society" (p. 75). What sources scholars have in order to gauge the workings of the past are due sometimes to bad luck (fire), sometimes to good luck (dry climate), and sometimes to social and political upheaval (no one has time to burn his papers if he's running for his life!). One reason so much work has been done on the social and religious history of *ancien régime* France is that the National Assembly seized the papers and other property of both aristocrats and the Roman Catholic Church during the French Revolution. Much of this was deposited in the Archives Nationales, where it is now available to scholars.

Contingency and catastrophe are only half the battle in the creation and maintenance of the archives. The other half is negotiating individual and group identities. More good luck for historians is when someone, whether relative or archivist, decides that some collection of documents, photographs, or artifacts is part of some sufficiently important framework that they should be preserved for posterity. Whereas "Great Man" history led to the collection of the papers and effects of monarchs, presidents, generals, and other public figures, beginning in the mid-twentieth century social historians instead emphasized dance hall tunes, broadsides, penny-dreadfuls, and those early daguerreotypes of serious-looking working-class folk. What gets into an archive depends on what individual archivists consider important, and that, in turn, depends on social frameworks that determine how a society views itself and on the kind of history it values. But the goals, plans, desires, and emotions of the entity also get into the act. As Bloch complains, some private subjects (corporations or families) are so concerned about their reputations, their current and future public identities, and how a third party could challenge their preferred self-narratives that "A good cataclysm suits our business better" than waiting for some persons of interest voluntarily to open their archives to scrutiny (Bloch, 1953/1964, p. 75).

Even objects that are (now) considered important to collective memory making—such as posters for public health education—are not guaranteed to turn up in an archives. Such "ephemera" have larger, more public audiences than family letters or business records, but even they must be evaluated within social frameworks in order to be considered important enough to be stored for possible future consolidation. Health practitioner and historian W.H. Helfand, for instance, has argued that public health posters are among the most effective mass media:

Posters have been a powerful force in shaping public opinion because propagandists have long known that visual impressions are extremely important in memory retention. Although people may forget a newspaper article, most remember a picture. A pamphlet or a newspaper can be thrown away, unread; the radio or television turned off; films or political meetings not attended. At some time or other, however, everyone notices messages when walking or driving, or sees posters on bulletin boards in offices, hospitals, clinics or pharmacies. (Helfand, 1990, p. 23)

What makes this medium so effective in its educational mission is that it should take no more than a single glance for a poster's message to be conveyed. Whether the viewer follows its advice is of course another matter, but unlike newspapers or even pamphlets, little effort is required on the part of the audience to get the point from a public health poster. Despite their effectiveness, the archival term "ephemera" is apt for these discursive tools, which are posted in high-traffic areas where the intended audience of "the public" has access to them. However, the secondary audience of scholars does not then have easy access to the posters, which often succumb to weather, vandalism, replacement, or neglect before samples can be "rescued" and furnished to a depository for safe keeping as items of material culture. The only reason such ephemera are archived now is because enough people have decided that public health narratives are important to societies' identities and histories, and they have been able to secure the resources to make the posters available for study (Helfand & Keister, 1990; Keister, 1990). We can conclude that greater attention increases an item's probability of being retained in collective working memory and later maintained as part of stable collective memory. Thus, attention is as much an integral part of collective as of individual short-term memory.

The Modern Collective Working Memory

The buffer, whether individual or collective, contains all of the labile memory items, and some of the downloaded stable memory items, possessed by its entity. In individuals, working memory is the part of the buffer that holds the items currently being processed by other cognitive subsystems. One subsystem that is central to consolidation is the relater (i.e., the hippocampus), which identifies and manages the associations between items that are abstracted by the generalizer in forming stable memory. The items in individual working memory include those to which an individual attends. By analogy, collective working memory is composed of the items in the collective memory buffer that are actively being processed by social subsystems, including the social relater (which we will call the "social hippocampus" in part III). As in individual memory, some of the contents of collective working memory are labile, such as news items, whereas others are derived from already consolidated collective memories, such as group narratives and social identities. Whether labile or stable, the items in collective working memory include those that have been tagged for attention through their activation of collective generalized memory or through the non-mnemonic exigencies of the social entity.

Again by analogy with individual memory, the items in collective working memory that are tagged for attention are more likely than the unattended items in the collective buffer to be consolidated. However, even on collective levels, certain items such as persistent features of the environment can be continually reintroduced into the buffer, whether or not they are attended, and can be consolidated along with the attended items in collective working memory. Still, it is easy to see how the items that a society actively considers are the items that eventually become integrated into its long-term collective memory. These consolidated memories may then recur to the collective buffer and influence attention and future consolidations, as the contents and meanings of collective memory change gradually over time and with the generations.

What the sketchpad is to individual working memory, the concept of the public sphere is to collective working memory. German sociologist Jürgen Habermas, in his classic study *The Structural Transformation of the Public Sphere* (1962/1989), asserts that the rising capitalist bourgeoisie created a "public sphere" in the early eighteenth century as a new area of influence opposed to the absolutist state (the political sphere) on the one hand and to the family (the domestic sphere) on the other. This space could be physical, as in taverns or coffeehouses, or more commonly—and in his view, more importantly—a discursive space for private individuals to "assemble" as peers and discuss issues of civic interest through talk and especially writing.

According to Habermas, discursive space is a precondition for democracy and part of the liberal narrative of progress from authoritarian oppression to free public communion. In that the public sphere facilitates "contest and negotiation" it is in part analogous to the relationality element (i.e., the relater) that we will discuss in detail in the next chapter. In that the public sphere makes items available for group consideration, it is also analogous to the buffering element, and specifically to the working memory component of the buffer. As for individual working memory, collective attention can shift among available items in the public sphere.

The News Directs Public Attention

A pervasive example of collective working memory is mass media, which exhibits characteristics of working memory that include attentional tagging and the recursive influence of stable frameworks. On a purely objective level, one might argue that the news media, in choosing which stories to report, should prioritize events according to their information content. Technically, the information content of an event goes up as the probability of its occurrence goes down (Shannon & Weaver, 1949). According to an information-content scheme, the story documenting that the sun rose this morning would get a very low priority, but the report that China has ceded Táiwān to the United States would be front-page news. Indeed, unexpectedness is an important factor in determining newsworthiness, but so are more subjective factors such as negativity, reference to elite persons or nations, and entertainment value (Galtung & Ruge, 1965; Harcup & O'Neill, 2001; Vasterman, 2005). The subjective factors depend on the recursion of stable collective memory: negativity, status, and entertainment value are only meaningful in the context of the shared narratives and social frameworks of the intended audience. Those news services that key into the social frameworks of their audience will be more successful at directing attention to their news (Winerip, 1998) and at influencing collective memory consolidation.

In individuals, attention is linked to working memory because most cognitive processes act on items to which an individual attends. Attention is also linked to individual memory consolidation through the process of modulation, by which an attended item can be selected for verbatim storage in long-term memory. Modulated items from the buffer can be taken up "whole" by the generalizer and fit into generalized frameworks without losing their specific characteristics (see also chapter 7). This form of modulation also occurs on collective levels, as illustrated using the example of a famous baby.

Americans who were adults in the 1980s probably remember the saga of Baby Jessica, the media event that developed after 18-month-old Jessica McClure fell 22 feet down an 8-inch-diameter pipe in her aunt's backyard in Midland, Texas, on October 14, 1987. Media worldwide covered the 58-hour rescue effort, and an ABC made-for-television movie, *Everybody's Baby: The Rescue of Jessica McClure* (1989), and a *Rescue 911* episode followed. Scott Shaw, working for the *Odessa American*, won the 1988 Pulitzer Prize for Spot News Photography for his work on the scene. At the time, a whopping 69% of Pew Research Center survey respondents followed the story "very closely"; by comparison, 80% very closely followed the *Challenger* disaster in 1986, whereas "only" 52% followed the *Exxon Valdez* accident in 1989 (Parker & Deane, 1997).

The intense collective attention focused on this story created a stable trace of it in consolidated collective American memory, which Americans continue to use for collective remembering. For instance, *Ladies Home Journal* carried exclusive updates on the 10th and 15th anniversaries, and *USA Today* named Jessica McClure the 22nd out of 25 "Lives of indelible impact" in May 2007 (Koch, 2007). MSNBC had sole coverage of the 20th anniversary, including the first live interview with Jessica herself (Celizic, 2007). This episode and its repercussions illustrate how enhanced collective attention can cause modulation of a selected event that leads to the consolidation of a national stable memory for that specific event. It also illustrates how social frameworks can direct collective attention: The nation's attention was riveted to this classic story of an innocent young victim and tireless heroes *because* American social frameworks already support such narratives of danger and rescue. This example illustrates how existing stable memories, by modulating the attention paid to certain events in the collective memory buffer, recur to influence the process of collective memory consolidation.

(Counter)memorialization and the Presence of Absence

Another example of collective working memory in action is the Holocaust countermemorial debate in Germany. Historians of art and culture now accept that even the hardest stone monument cannot fix collective memory into an immutable form, because the monument is only a tool for remembering. As Austrian writer Robert Musil (1880–1942) wrote in the 1920s, only partly in jest, "monuments are so conspicuously inconspicuous. There is nothing in this world as invisible as a monument" (Musil, 1936/1987, p. 61). Starting in the 1980s, as part of a broader discussion about Germany's "working through" its past, there developed a movement to discourage the public from trying to displace memory and guilt onto static and easily forgettable memorial plaques or statues. Following the spirit of Theodor Adorno's famous dictum, "To write poetry after Auschwitz is barbaric," countermemorialists have especially shunned figurative memorials. How can one represent the irrepresentable or speak the unspeakable (Trezise, 2001)? Countermemorialism proponents seek new ways to consolidate collective memory without the petrifaction and loss of meaning that seem to follow memorialization. They worry that the act of erecting a monument serves as a catharsis of pent up tension and not as a stimulus for continued remembering. Musil blamed their very stability, their ever-presentness, for monuments' seeming invisibility. Either way, monuments seem to consign collective memories to the void of the forgotten.

It is precisely traditional monuments' uncanny ability to "repel attention," as Musil put it, that bothers countermemorialists (Young, 1993). What a countermemorial therefore tries to do is to reintroduce continuously the consolidated past back into collective working memory. Whatever a monument's physical solidity, countermemorialists attempt to stretch out the memory's discursive lability and to keep the collective's attention on it. Probably the most (in)famous countermemorial is the Harburg Monument against Fascism, War and Violence—and for Peace and Human Rights, designed by Jochen Gerz and Esther Shalev-Gerz and inaugurated in the Hamburg suburb of Harburg on October 10, 1986 (Lupu, 2003). The Harburg installation consisted of a 12-meter-tall stele of hollow aluminum covered in lead. Visitors were invited to inscribe their names in the soft lead as a promise to work against fascism, war, and violence (and for peace and human rights). When the bottom of the stele had been covered in writing, it was lowered into the monument's base 40 centimeters, bringing a fresh area of writing surface into reach. Eight times the stele was lowered, until finally, on November 10, 1993, the stele was sunk completely into its stone base. All that remains visible of the vanished monument is what visitors can glimpse through two windows in the sides of the base.

Explains political scientist Noam Lupu, "the Gerzes envisioned the monument as a work-in-progress, a blank slate on which to reflect public sentiment back onto individuals. . . . The monument would provoke Harburg citizens to do the work of *Denkmal-Arbeit* ('memorial work') themselves, to themselves become the monument, and to act against the resurgence of fascism and intolerance" (Lupu, 2003, p. 145). The counter-memorialist intention was that, "By 'committing' themselves to vigilance, the citizens of Harburg would turn memory into action, would themselves become the agents of *Denkmal-Arbeit*" (p. 138). With their countermemo-rial, it appears that the Gerzes attempted to invert the usual progression from short-term to long-term collective memory. They started off where many consolidation processes end, with an imposing, 12-meter-tall monu-ment, a memory structure that most observers would associate with already consolidated collective memory. By causing this monolith gradually but conspicuously to disappear, and by giving rise to active participation (inscribing of names) and discussion, they reintroduced the subject of the Holocaust back into collective working memory. Rather than end with a potentially forgettable item of material culture, they converted it into discourse, which, they hoped, would lead to actions chosen to combat continued crimes against humanity. They also hoped it would lead to collective consolidation of an inclusive, tolerant narrative.

Lupu calls attention to a paradox in the concept of countermemorializa-tion. This and other countermemorials frequently use negative space to mark "the presence of absence" (see especially Young, 1993, 2000). The Gerzes' vanishing stele shares something in common with another countermemorial, Horst Hoheisel's Aschrott Fountain in the City Square in Kassel. In 1908, Jewish businessman Sigmund Aschrott had donated a 40-foot-tall pyramid-shaped fountain to the city. The Nazis destroyed it in 1939, and after the Second World War, flowers were planted in its place. With no small irony, the locals called the garden "Aschrott's Grave," but eventually no one remembered who Aschrott was. In 1984, Hoheisal replaced the flower bed with a new fountain, a 12-meter-deep funnel into which the water disappears. The funnel is an inverted pyramid, and Hoheisel hoped by its extreme negative presence to highlight both the absence of the fountain's namesake (and his entire religious community) and the decades of silence about the violence against both fountain and community. By removing the innocuous presence of the flower bed from the material, public sphere, Hoheisal was trying to incite remembering in the discursive sphere.

Whatever their creators' intentions, Lupu argues that such "negative monuments" feed into traditional understandings of the Holocaust. Judging from comments made at the opening of these and similar memorials, the public expected appropriately somber places of reflection; the Aschrott Fountain in particular "reinforced preexisting representations of the Holocaust meta-narrative, reviving old metaphors: the Holocaust as abyss, a dark history in the depths of German consciousness" (Lupu, 2003, p. 150). In other words, Hoheisel's avant-garde countermemorial had done nothing more than confirm what people already believed about the Holocaust: Instead of having to confront the possibility of violence or oppression among themselves in the present, visitors tended to remember how bad things were "back then."

What becomes obvious is that even the "presence of absence" is a presence, and presence is the hallmark of a traditional monument, the very thing against which countermemorialists aligned themselves. Lupu has found that the audiences for these two countermemorials did not "get" it. Germans in Harburg and in Kassel interpreted the "negative monuments" as symbols of loss; there is little indication that they filled the discursive sphere with remembering in response to the voiding of the material sphere. Thus, we see the importance of material culture for the content of discursive space (e.g., Seremetakis, 1996). The material culture of memory—monuments, books, pictures—may be more necessary for both the consolidation and the subsequent remembering of collective memories than some countermemorialists would like to admit. Although the physical characteristics of objects do not determine whether they serve short-term or long-term memory, and although remembering on either timescale is a

dynamic process, both short-term and long-term memory require a physical substrate.

Conclusion

The buffer can be thought of as the "intake" element in the memory consolidation process. It holds new (labile) memories and also holds old
(stable) memories that are downloaded to it from the generalizer. It serves
as a "workbench" on which memory items are made available "to hand"
for selection, relation, and generalization. That part of the buffer in which
items are actively being processed corresponds to working memory, and
the items in working memory are usually those to which attention is paid.
Attention makes labile items more likely to be selected for eventual consolidation. Attention can also mark items in the buffer for verbatim consolidation, in which, through the process of modulation, their specific
characteristics are preserved while their association with more generalized
knowledge is also registered. Far from being an indiscriminate collection
of all labile potential memories, the contents of the buffer are regulated
both by recurrent, stable memory from the generalizer and by the goals,
plans, desires, and emotions of the entity. Through their influence on the
contents of the buffer, the generalizer and the entity influence the consolidation process.

In this chapter, we encountered buffering concepts such as the "phonological loop" and the "visuospatial sketchpad," which are thought to
hold items over the short term in individual memory. Models containing
these or similar buffering elements achieve greater power to explain the
real working memories of individuals when their contents are subject to
the whims of a central executive or, more realistically, to the influences of
already consolidated, long-term memory. The individual buffer is therefore
best seen as an actively managed collection of memory items, and this
view extends equally well to the collective buffer. We saw how the postmodern concept of "the archive" corresponds to the buffer in our three-
in-one model, where the influences of the generalizer and entity correspond,
respectively, to the forces that social frameworks and institutions exert on
collective levels. We described the importance of collective attention, as
directed by societal phenomena such as mass media, in marking items for
collective consolidation. We also described the countermemorial effort, by
which artists attempt to make stable memory labile again, but were led to
conclude that memory structures, whether serving short-term or long-term
memory, are necessary after all.

The buffer that emerges from our examination is an actively managed repository, but one that is still intended as a short-term store from which items will be selected and related for transformation into long-term memory. This chapter has acknowledged the substantial influence on the contents of individual and collective buffers that is exerted by established frameworks and subjective biases, but taken to an extreme these influences would render the buffer useless to the remembering entity. Ultimately, memory must serve the entity's need to survive, so it must consolidate at least some knowledge that is accurate and objective in order for the entity to adapt in a changing environment.

Even the most autocratic entity must allow intake into its buffer of new items that reflect not its preexisting frameworks and biases but the actual structure of the world. It may be useful, for example, for an entity to consolidate frameworks that emphasize its importance, even if exaggerated, but it is critical for it to consolidate knowledge constructs that represent certain aspects of the world as it actually is. Just as you need to remember that a red light means "stop," no matter how important the meeting you are rushing to may be, so all entities must mix some reality in with their goals and desires else they fail to realize and satisfy them. Our exploration in this chapter leads us to conclude that established frameworks and subjective biases can determine which items are tagged for attention in individual and collective memory buffers, and can also influence buffer contents, but the environment is and must always be the original source of new, labile, potential memory items. From the buffers, items are selected and related to shape and reshape usable knowledge constructs. It is to the processes of selection and relation that we turn in the next chapter.

6 Selection and Relationality

The second element of our "three-in-one" consolidation model that we will discuss is the selector/relater, whose function is to select items from buffered storage and establish meaningful relationships between them. The selector/relater is the central element among the three core elements of our consolidation model (buffer, selector/relater, and generalizer). It is the core element that is arguably the most strongly influenced by the consolidating entity (the fourth element, which is an encompassing meta-element). The selector/relater is also strongly influenced by recursive feedback from the generalizer. The central importance of the selector/relater to the consolidation process is apparent when one realizes that the content of stable memory is not just a collection of isolated facts. Rather, the products of the consolidation process are efficiently organized webs of associations and generalized knowledge constructs that are derived from the systems of relationships the selector/relater establishes.

Recent research testifies to the critical importance of the selector/relater to individual memory consolidation and opens a new perspective on classic findings. The era of neuropsychological memory research began with the famous patient H.M., who exhibited retrograde and anterograde amnesia after bilateral removal of his temporal lobes, which contain the hippocampi (there are two hippocampi, one on each side of the brain; see chapter 2). Recent work indicates that the main functions of the hippocampus (referring to both hippocampi) in cognition and in individual memory formation are to establish, represent, and access systems of relationships. As the case of H.M. showed, removal of the hippocampus, and with it the process of relationality, has devastating consequences for declarative memory formation in individuals. By analogy, the same critical role is played by the relation-makers in society (including opinion leaders in journalism, academics, politics, etc.) who function as its "social hippocampus."

At its most basic, relationality makes memory items (mental representations of objects, facts, events, ideas, etc.) meaningful because of their objective connections to other items. For example, the events and persons represented in a system of relationships that will ultimately consolidate in narrative form will be related through temporality and emplotment. In contrast, the facts and ideas included in a system of relationships that will ultimately consolidate as a scientific paradigm are related through correlation, causation, and the correspondence between theory and observation. In selecting potential memories from buffered storage for consolidation, the relationality element may link new (unconsolidated) as well as old (already consolidated) memory items, because the generalizer can recursively download old items to the buffer where they can mix with new items. The selection and relation process is also subject to control by the goals, desires, plans, and other non-mnemonic properties of the entity. The result is an integrated system of relationships that the generalizer abstracts and stores as stable memory (see chapter 7). Because of the influence of the entity, this stable memory can be subjective in ways that make it more acceptable to the entity, but intolerable inaccuracies eventually can be corrected because consolidation is an ongoing process that continually reshapes stable memory.

The exception to the rule of integration is the phenomenon known as "modulation," which occurs when a "tag" of importance is attached to a memory item. Modulation can be thought of as a stronger form of attention, and one that is often associated with emotion. As with attention, modulation can occur automatically, as when new memory items in the buffer activate old memory items in the generalizer (see chapter 5). Alternatively, modulation can occur as the result of control by the entity, according to its various motivations. Although the consolidation process firmly establishes their relationships with other items, those items tagged through modulation can be taken up whole into stable memory without losing their specific characteristics, even as the particulars of most other items are lost through generalization. In this way, you vividly remember your "first fish," and it is related to your general knowledge of fishing, though it will forever occupy a "special place" in your memory.

Emotion and memory have a two-way relationship: emotion can modulate a particular memory item and influence the way it is represented in stable memory, but an event that triggers certain "special" stable memories can also arouse emotion (Finkenauer et al., 1998). Memories associated with high emotion tend to be vivid, but the vividness of a memory can be completely independent from its accuracy, as we will see in the next chapter. And just because a modulated memory is specific does not imply that its relationships are confined to one context. On the contrary, an item can arouse emotion *because* of its connections to memory items in multiple contexts; your first fish is as likely to be related to significant persons in your life as to your general knowledge of fishing. The result of modulation on the consolidation process is the production of a stable memory organization in which specific items are imbedded within more generalized knowledge constructs. Thus, stable memory is both specific and general.

The individual memory phenomenon of relationality is known in neuroscience as *binding* (Zimmer et al., 2006). The first half of this chapter reviews the development of relationality in neuroscience by discussing the history, anatomic substrates, and proposed functional mechanisms of binding, particularly as these concern the hippocampus and the medial temporal lobe. It also reviews the literature in cognitive science, which has long used the idea of a relational network to describe memory and language processing. Working on the conceptual level, we also apply to individual memory a concept popular in the fields of art and art history, *polysemy* (from the Greek for "many meanings"). We conclude this section by emphasizing that interrelationships between items are the very essence of memory.

The second half of this chapter concerns characteristics that the collective selector/relater shares with its individual counterpart. As in individual memory, emotion also has a two-way relationship with collective memory: Emotion modulates the selection and relation of specific memory items, and subsequent activation of those memory items arouses emotion (Cohn et al., 2004; Lambert et al., 2009). Perhaps because it is a macrolevel process—and probably also because context is already so important for work in the humanities—the collective selector/relater easily exhibits both the recursive effects of already consolidated memory and the non-mnemonic influences of the consolidating entity. We begin the second half by exploring the collective establishment of relationships between interpretations and emotion-laden memory items with reference to journalistic coverage of two mid-twentieth-century American events: the Kennedy assassination and Watergate. Journalists are not the only relation-makers in American society, so we will also explore the concept of a "social hippocampus" and the possibility that a collective contains many, perhaps overlapping, "social hippocampi."

Because there are many groups in society, a single object or event is almost always polysemic in that it can have many meanings in relation to the narratives (in generalized memory) and goals (and other entity properties) of the various groups. Precisely because there are so many different ways to interpret an event, a certain amount of "contest and negotiation" frequently accompanies a collective's efforts to consolidate a single collective memory (Wertsch, 2002, 2009). In some cases, intergroup striving leads to the formation of a more accurate and complete collective memory, but in other cases it leads to a fractionation in which the various groups consolidate their own separate versions of events, whereas the larger society is left with a meager and insipid account.

We will explore such a fractionation with the example of the heated debate surrounding the design of the *Enola Gay* exhibit at the National Air and Space Museum in Washington, D.C. In this matter of American history and self-image, the efforts of various stake-holding groups to establish the "proper" relationships between labile facts clashed with the stable but conflicting social frameworks and entity properties of those same groups. Divergent narratives can also lead different groups to form different systems of relationships from the same set of facts, leading to the consolidation of conflicting collective memories. We end the chapter by presenting (and contrasting) two museum examples, one from the Greek side and the other from the Turkish side of the island of Cyprus, which illustrate the effects of divergent narratives and collective goals, desires, and emotions on collective relationality.

Relationality in Individual Memory

If relationality is, most generally, the linking (or binding) of two or more memory items, then this process can be seen on many levels in the brain. At a very low perceptual level, for instance, the brain combines the features of individual objects (color, shape, separation) into mental representations of the objects themselves. To distinguish one object, event, or item from another, the brain must represent which features are related to which items. A representation must contain the features of an object but also its separation from other objects (Treisman, 1996). What is it? What is it not? (What is not it?) Researchers in feature binding attempt to describe how the brain uses the process of binding to create multilevel representations. For example, the representation of an individual flower involves binding together its shape, color, texture, and other features; the representation of a bouquet involves binding together the representations of individual flowers; the representation of a flower shop involves binding together the representations of many separate bouquets. Feature binding is involved in a low-level recognition such as red + octagon + letters = "an object that is not a car" or a high-level memory such as stop sign + car + terror = "the time I was almost hit by a car that ran a stop sign." Feature binding is an important part of perception, because it converts isolated percepts into a complete, meaningful picture.

Relational binding, which occurs at a higher level than feature binding, is a powerful process because it imparts meaning(s) to the representations it creates, not just for objects but also for concepts. For instance, to understand the meaning of a word in natural (i.e., human) language, it is necessary to relate the word to the context in which it is used. The connections formed between a word and its contextual elements bind that word into a network of meanings. By "surfing" the connections between each word, it is possible to reconstruct not just isolated items but rich and meaningful contexts in which these words were previously experienced (Trinkler et al., 2006). These networks are also responsible for the ability of many individual words to evoke more than one meaning: they create polysemy.

Linguist Noam Chomsky's idea of the "deep structure" of phrases or sentences relies on such networks of meanings. Chomsky (1965) understands language as not merely a string of words (surface structure) but as the meaning produced by the words (and their related concepts) in particular relationships to each other (deep structure). Words are thus in relationship with their many definitions and, in phrases or sentences, with each other. Consider a classic double entendre such as the headline, "Prostitutes Appeal to Pope." The surface structure is nothing more than the sequence of words. But the deep structure contains the many meanings of that particular sequence, and which meaning is imparted depends heavily upon the context. Read on the front page of a reputable newspaper, one could assume a group of sex workers had asked the pope to consider an issue of importance to them, such as their human rights. Read in a tabloid magazine, one would assume the pontiff had a hankering for "ladies of the night."

Experiments in experimental psychology clearly show the influence of prior knowledge from the generalizer on the relationality process. A canonical example is the Deese–Roediger–McDermott (DRM) experimental paradigm. It all began in 1894 when, as part of an unrelated memory experiment, American psychologist E.A. Kirkpatrick noted that students would often testify to remembering words that were not on a previously presented list but were instead somehow associated with the words on the list. As Kirkpatrick put it, "things suggested to a person by an experience may be honestly reported by him as part of the experience" (Kirkpatrick, 1894, p. 608). Psychologist James Deese expanded on this work in 1959 to create an experimental paradigm to test such persistent, inaccurate experimental reporting, and in 1995 cognitive psychologists Henry Roediger and

Kathleen McDermott revived the procedure as a means of testing memory. The resulting DRM paradigm involves subjects listening to a long list of related words, for example:

bed, rest, awake, tired, dream, wake, snooze, blanket,
doze, snore, slumber, nap, peace, yawn, drowsy.

Missing from this list is a single, strongly associated "lure" word. Can you guess what it is?[1] When subjects are asked to recall as many words from the list as possible, the lure word has an extremely high rate of false-positive response. Depending on the number of related words and their degree of relation, the rate of false remembering approaches 55%, and the rate of false recognition (incorrectly asserting that the lure word appeared on the list) approaches 80%—as high as the rate of correct recognition (Roediger & McDermott, 1995).

   This experiment demonstrates what is known as "spreading activation in a semantic network." The list items activate not just their own stable representations but also all the words that already have associational links with them. Although *sleep* is never primarily activated by being heard, it is secondarily activated 15 times, since all of the items on the above list are related to sleep. In the context of our model, the words on the list enter the buffer as new items, and the relater relates them to each other and to old items. Through this process, they activate the word *sleep* in the generalizer, which then downloads *sleep* to the buffer, where it joins and is related with the other words on the list. Now, when "the list of words" is consolidated, the lure word *sleep* is included, even though it was not a member of the originally presented list. When asked to recall the items on "the list of words," subjects recite all the words they can remember, which very likely will include the lure word. Thus, the backward arrow in our model from the generalizer to the selector/relater is justified, because the recursion of old knowledge and old items strongly influences how new labile items are selected and related to each other and to old items that are already represented in stable memory.

Anatomy of a Memory

After H.M., and after many confirmatory observations, neuropsychologists thought that all of memory consolidation occurred in a single brain structure, the medial temporal lobe (MTL, which contains the hippocampus;

1. *Sleep*, of course!

see figure 2.4) as damage to it produces profound memory deficits (Cohen & Squire, 1981; Markowitsch & Pritzel, 1985; Cohen & Eichenbaum, 1993; Squire & Alvarez, 1995). Neuropsychologists have since narrowed the functions of the MTL to just part of the consolidation process: relational learning. What functional role does the MTL serve, such that its removal (in patients such as H.M.) would prevent the formation of new memories and abolish a certain proportion of the old?

Drawing from the literature on retrograde amnesia in patients and animals (see chapter 2), in addition to anatomic findings, Larry Squire, Neal Cohen, and Joyce Zouzounis (1984) (see also Alvarez & Squire, 1994) proposed that the MTL acts as a temporary facilitator of the interactions between neocortical memory storage regions that eventually can be activated independently from the MTL. The process can be described in terms of labile (short-term) and stable (long-term) declarative memory: Labile representations in the MTL drive neocortical representations of a new set of memory items. Eventually, connections within the neocortex will "learn" *most* of the declarative memories that had been represented first by the MTL and then by the connections between the MTL and neocortex (there is some loss). At this point, the MTL is no longer needed to drive the neocortical representations and the declarative memories can be considered to have consolidated as stable neocortical memories. As H.M.'s case demonstrates, the entire medial temporal structure could be removed without disturbing the now-consolidated memories. What would be disturbed, however, would be those memories whose connections the MTL and neocortex were still rehearsing, and retrograde amnesia would result.

Research over the past decade has refined our understanding of the MTL. Investigators have determined that the specific role of the MTL is to guide relational learning and that this learning is responsible for the long time-course of consolidation as revealed in cases of retrograde amnesia. In light of new findings, focus has shifted from the MTL as an organ of consolidation to the MTL as an apparatus of *relationality*, one that connects selected items together to form a representation of the world using the relational binding process discussed earlier (Cohen et al., 1997, 1999). According to this hypothesis, MTL activity ultimately creates in the neocortex a stable network of associations that represents the relationships between declarative memory items and that can be activated independently of the MTL.

This conclusion fits well with findings about the MTL from animal studies (O'Keefe & Nadel, 1978; McDonald & White, 1993). For instance, O'Keefe and Nadel found that removal of a rat's hippocampus (a component of the MTL) causes an inability to complete tasks that require processing of relationships between specific memory items and contextual clues (such as navigating a maze based on the position of objects in the room). Howard Eichenbaum and co-workers have observed similar deficits in the ability of hippocampectomized rats to perform tasks that require them to process associations between presented scents (Bunsey & Eichenbaum, 1993; Eichenbaum & Otto, 1993). Human amnesics with damage to the hippocampus similarly have trouble learning complex relationships between items, but they are not impaired at identifying specific items already consolidated and stored in stable memory (Hannula et al., 2006a, 2006b). Neuroscientists have concluded that context is implicated in the consolidation of memory and that the hippocampus is necessary for the processing of associations.

Eichenbaum reviewed the full implications of this theory in his 2006 article "Memory Binding in Hippocampal Networks." He asserts that the hippocampus is not responsible merely for the consolidation of memories in their contexts but that the hippocampus is responsible for *all* relational learning. Drawing on work by computational modelers of hippocampal dynamics, Eichenbaum describes the hippocampus as a system that builds relational links between stored memories according to their approximate temporal order, spatial relationships, or common features. These "hippocampal objects" can represent the interrelatedness of the memory items that compose a particular situation or state of affairs.

Such hippocampal objects could be understood as episodic memories: complex declarative memories characterized as interrelated sets of simpler declarative memories including facts and events (Tulving, 1972; Rolls & Kesner, 2006). Once an episode has been established by the creation of a set of links between items, this hippocampal object can be used as a substrate for generalization, the process that extracts the structure inherent in the links within and between many such episodes for use in novel contexts. Hippocampal relationality is revealed by experiments demonstrating that rats can learn transitive associations between stimuli never presented together (e.g., Bunsey & Eichenbaum, 1993). Given the two pairs of scents A–B and B–C, rats will later correctly choose the novel pairing A–C, rather than a pairing with a lure scent (e.g., A–X). Rats can also learn symmetric relationships (e.g., if B–C then C–B and not C–Y). The sequences can grow quite long, and the tasks can become correspondingly more difficult, but even rats, as Eichenbaum (2006) found, can learn and process scent chains up to six items long! This learning cannot occur in hippocampectomized rats. This work (reviewed in Eichenbaum, 2006) demonstrates that rats can represent the ways in which a set of scents are associated and then "surf" this web to extract meaningful relationships (e.g., if A–B and B–C then A–C and not A–X), provided they have intact hippocampi. These results provide strong experimental support for the idea that the function of the hippocampus is to represent relationships between memory items and to flexibly access those representations.

What Is the Meaning of It All?

However new in neuroscience, relational binding is nevertheless an old concept in cognitive science. Researchers in computational linguistics, for instance, have used relationality in their conceptualization of semantic networks. As far back as 1969, M. Ross Quillian, now a political scientist, designed a "teachable language comprehender" that learned the meanings of words (rather than the sounds of them) by linking the definitions of words that appear together in context (Quillian, 1968, 1969). This program developed a complex tree of associations that it could query in order to explain the content of a given input. For example, given the statement "The spouse is suing for custody," the program might divine the meaning as "there is a client who employs a lawyer, has children, is divorced, and is attempting to gain custody of them through legal action." It could perform this task because it could not only climb the tree of associations but also descend to access other limbs and discern which of the multiple possibilities for concepts such as "spouse" and "custody" is relevant to the situation.

The relational aspect of memory is self-evident when one considers that many different cues can trigger the same memory item. Helen Williams and Martin Conway (2009) have proposed the concept of autobiographical memory networks to account for the relational aspect of the memories a person has about his or her own life. They posit that an individual has not one but many autobiographical networks that interconnect, so that memory items stored in one network can be recalled by associations that have been activated in another network. As an example, they point out that "recalling a specific episodic memory of meeting a girlfriend for the first time could be cued through discussion of that girlfriend, discussion of girlfriends in general, discussion of the town in which you met, or discussion of the particular bar in that town where you were working on the night that you met her" (Williams & Conway, 2009). The memory networks in this example, which combine specific items (e.g., "the particular bar") and general concepts (e.g., "girlfriends in general"), are already consolidated. We will discuss the roles of specialization and generalization in the process of memory consolidation in chapter 7. Our emphasis in this chapter is on the process of relationality that establishes the networks of relationships from which the generalizer/specializer draws as it lays down stable memory.

The idea that the same memory item can be an element of several different memory networks suggests that memory items are polysemic in individual as well as in collective memory. Although memory polysemy is clear for individual humans, what about individual rats? After all, how many meanings could learning a maze possibly have for a laboratory rat? Yet the idea of memory polysemy is useful in explaining a seeming memory paradox. In what has been termed memory "reconsolidation," an already consolidated memory item is thought to become labile again when it is recalled, because it seems to be forgotten if recall is accompanied by certain experimental manipulations (Riccio et al., 2006). To take an example, a rat that has learned a maze to find a food reward seems to forget the maze if, when reintroduced to the maze, it receives electric shock rather than food upon reaching the end. In his refutation of reconsolidation, James McGaugh (1972) speculates that the rat may not "forget" the maze after having been shocked; rather, it now also associates completing the maze with the unpleasant experience of shock. This makes "completing the maze" a polysemic memory item whose meaning had been, and could again be, positive (food!) but is currently negative (shock!). When the shock is replaced by food, the rat with this possibly polysemic representation learns (or relearns) the maze much more quickly than a naïve rat, because the experienced rat only needs to realize that "completing the maze" has regained its former meaning. (We will discuss the controversy over reconsolidation and revisit this example of polysemy in chapter 12.)

Thus, relationality in individual memory serves to impart meaning to stimuli by connecting items together. In the words of William James, quoted in part earlier:

In mental terms, *the more other facts a fact is associated with in the mind, the better possession of it our memory retains.* Each of its associates becomes a hook to which it hangs, a means to fish it up by when sunk beneath the surface. Together they form a network of attachments by which it is woven into the entire tissue of our thought. (James, 1890, p. 662, emphasis in original)

Relationality allows for the establishment, representation, and processing of relationships among items and between items and contexts, creating networks of meaning. As we will see in the following chapter on generalization and specialization, abstractions from established systems of relationships provide coherent frameworks in which specific details can be embedded. These exemplars and details are the "hooks" of which James writes. When new things can be attached to old hooks, learning them is so much easier than when there are no hooks available. Patients like H.M. demonstrate how important to consolidation is the process of relationality, for they lack the ability both to connect new items to each other (i.e., a face and a name) and new items to old ones (i.e., daily events into a continuous autobiographical narrative). As H.M. himself poignantly put it, "Every day is alone in itself, whatever enjoyment I've had, and whatever sorrow I've had" (Milner et al., 1968).

Relationality in Collective Memory

Of all the operational components of the consolidation process, the parallel between relationality on individual and collective levels is particularly strong, even though neuroscientists have only recently begun to embrace what for humanists has long been a central concept. We consider an analogy between the MTL (and especially the hippocampus) in individual brains and "the relaters" (groups of individuals) in society that, by virtue of their role in establishing relationships between potential items of stable collective memory, serve as social hippocampi. We know that the process of consolidation can take years in individuals, and we can infer that the reason for the delay concerns the need to reconcile new memory items with each other and with old items, and we can image that this reconciliation could involve competitive interactions occurring in the brain. Although competitive interactions between neurons are known to be involved in many neurophysiologic processes, they have yet to be explored in the context of memory formation (but see chapter 12 for a discussion of this possibility). Being locked as they are inside the skull, the activities of neurons require great effort and ingenuity to observe. Not so the activities of the various relaters in society. They live their competition out loud as they strive to establish the webs of associations between events from which consolidated collective memory will be distilled. Indeed, contest and negotiation permeates the politics of collective memory formation both within and between groups, as social hippocampi endeavor to determine the features of historical events as well as their meaning(s) (e.g., Wertsch, 2002, 2009).

Feature binding, like its higher-level cousin relational binding, is easier to observe on collective levels than on the individual level because the process of collective binding takes place in the forum of open discourse. This accessibility reveals not only the process itself but also the unavoidable influence of already consolidated collective memory and of the goals, desires, and emotions of the collective entity. Historians, for instance, have long debated the "shapes" of past events, and defining what an event "is" (or "was") in relation to other events can be complicated. There may be no dispute concerning the occurrence of the event itself, but it is necessary to place the event in a context, determine a beginning and end for the event, its major protagonists and antagonists, a location, and so on. Historians once thought (and many laypeople still do) that there is only one, true way to bind the features of an event (e.g., Crew, 1996). In recent decades, however, with the development of women's and area studies, cultural history, queer theory, and other approaches to the study of collectives and their pasts, it has become evident that even determining the basic features of past events involves the social frameworks and animating impulses of the historians or other "relaters" engaged in making those determinations.

Revolutions are an excellent case in point. When considering the "Revolution of 1848" in relation to Germany, for instance, a historian would have to decide how broad a geographical scope to take (local, pan-German, European, global?) and a chronology. How early is the first important event? The barricades in Paris in February 1848? The unrest over the famine since 1847? The French, American, or English Revolutions (1789, 1776, 1666)? When does the Revolution of 1848 end? With the recapturing of Berlin by the monarchists in November 1848? With the "crowning" achievement of the counterrevolution, President of the Second Republic Charles Louis-Napoléon Bonaparte's establishment of the Second Empire in 1851? With the belated triumph of democracy in Germany in 1946?

Which of the possibilities a historian chooses depends, to a large extent, on the social framework of the particular group to which the historian belongs. For instance, a radical, democratic view currently popular among some scholars of German history pushes the boundaries of the revolution as far as possible. According to this perspective, the revolution encompasses revolts in Switzerland, Germany, Denmark, Transylvania, Austria, Sicily, the Italian peninsula, France, and French dependencies in the Caribbean. Active revolution thus extends from late 1847 to summer 1849. Moreover, because the goal is the democratization of Europe, the revolution's meaning spans the history of modern democracy from the French Revolution of 1789 to the current Federal Republic of Germany (Tacke,

2000). In contrast, at least until the second half of the twentieth century, conservatives and liberals alike denied that any true revolution had happened at all. Conservatives, because of their continued, nominal hold on political power, connected the features of the supposed revolution into as much of a nonevent as possible. Liberals associated revolution with political change, as had in fact occurred in France around 1789, so they similarly attached a meaning of failure to the supposed revolution, mostly because no unified German nation-state emerged. These and other possible definitions of the Revolution of 1848 between the two extremes of "global revolution" and "no revolution" depend on one's social framework, group narrative, or worldview.

Compared with feature binding, relational binding involves even more contest and negotiation on collective levels, as illustrated by the "critical incidents" (Gerbner, 1973) of mid-twentieth-century American life and politics. In *Covering the Body*, journalist and scholar Barbie Zelizer (1992) makes two points about the journalism profession's involvement in collective American memory of the assassination of President Kennedy. The first point is that journalists have insisted on their priority in both describing and interpreting the events on and around November 22, 1963. For almost three decades, they crowded out both historians, the traditional arbiters of meaning in a modern nation's autobiography, and independent critics, to be what one could call the "social hippocampus" for this collective memory. They claimed they had the duty and the right to establish the relationships for this critical event.

Journalists also offered to provide the consolidated memory of the event for American society, although Americans as a whole (if we can even consider such a whole) have yet to reach a consensus concerning the assassination. Zelizer's second point is that journalists themselves had more or less settled on a single understanding, that of "the end of Camelot." Kennedy's death was supposed to evoke images of a charismatic leader of a prosperous country and of a father to a beautiful family whose fabulous life was tragically cut short. Part historian, part critic, Oliver Stone challenged this relation of "assassination—end of Camelot" in his film *JFK* (Ho et al., 1991), which was very poorly received by a journalism profession Zelizer characterizes as jealous and much more subjective than it claims to be. Stone's movie attempted to reinforce the paired association of "assassination—conspiracy" and strengthen the connection between the shooting and the web of conspiracy theories in American culture. Which association would win out as the predominant American collective memory has thus been a matter of contest and negotiation.

Sociologist Michael Schudson (1992) approaches *Watergate in American Memory* in a similar way. In this case, journalists connected President Richard Nixon's cover-up of his involvement in the break-in at Democratic Party headquarters with "scandal," an interpretation reinforced by President Gerald Ford's pardon of Nixon (only criminals need to be pardoned) (e.g., Woodward, 1999). Historians, however, have fairly consistently related the whole incident to Nixon's political rivalries while in office (e.g., Kutler, 1990). According to them, his resignation signaled not national disgrace but the success of the balance of power in the federal government. These are just two instances of the phenomenon of relationality in collective memory, in which different groups in society confer different meanings on a single event.

These two "critical incidents" in American history (JFK's assassination and Watergate) are polysemic insofar as different groups in society gave each one multiple meanings. Any societal event is polysemic because "there are as many collective memories as there are groups and institutions in a society," so any event potentially has dozens of meanings within collective memory (Halbwachs & Coser, 1925/1992, p. 22). It follows that there is not one social hippocampus but many social hippocampi, because a "society" is not a single entity with a single collective memory but rather a set of groups, each with its own memory (Wertsch, 2002, 2009) and its own relation-makers. Thus, although there may not be a single social hippocampus in a collective, its function as a relation-maker is nevertheless present in various forms for the many subgroups that make up a society.

Relationality and Intergroup Interaction

Despite the multiple possible meanings of an event or idea, certain societal factions often proceed as if only one can be the "correct" interpretation. This leads to debates between groups espousing different and sometimes contradictory meanings, and the resulting contest and negotiation about what relations to draw between items in collective memory can become quite contentious. These intergroup conflicts illustrate a critical aspect of collective memory formation: Not only is the process of collective relationality influenced by the social frameworks and animating impulses operating within a group, but it can also be influenced by interactions between groups. We can think of these interactions as occurring at the level of the collective entities in question. An entity could modify its own memory consolidation process according to the outcome of an intergroup interaction, to the extent that the entity is convinced by the interaction. Often, through this mechanism, intergroup interaction succeeds in changing the system of relationships established by a group, and this leads to remodeling of its consolidated collective memory. To take a case in point, Native American groups have successfully changed the way American academics view and collectively remember their history and culture (Littlefield, 1992). Sometimes intergroup interactions fail to change the systems of relationships established by any of the interacting groups. This leaves a set of groups each with its own separate but potentially rich collective memory, whereas the larger society gets a bare, stripped-down version of events that seems to call out for further consolidation.

The object at the center of one recent public controversy over collective memory consolidation in the United States is the bomber, the *Enola Gay*, around which the curators at the National Air and Space Museum (NASM) had planned an exhibit for 1995 entitled *The Crossroads: The End of World War II, the Atomic Bomb and the Cold War*. Although the scientists and soldiers involved in the Manhattan Project and the bombings have often been portrayed as heroic and patriotic, in the NASM proposal, the good that was the end of the war was tempered by the suffering, death, and destruction that "Little Boy" and "Fat Man" wrought on Hiroshima and Nagasaki, and by the dawn of the atomic age. That collective memories are real forces in the workings of societies is revealed by the furor that erupted: veterans were concerned at least as much about the version of history that would be taught to younger Americans as about their own personal reputations; the antipathy of historians for anything that wafts of censorship led NASM director Martin Harwit to resign; and the U.S. Senate held hearings about the use of federal funds and influence. (One accusation was that the government institution dedicated to preserving aircraft was in fact discrediting airpower.)

This episode is usually cited in collective memory literature as a debate about the authority of participants in an event over its portrayal; as a worst-case scenario of censorship; and most frequently as the triumph of simplistic collective memory over nuanced history (e.g., Winter, 2009). We will return to the collective memory versus history debate in chapter 12. Here we see the episode as one of different groups in American society fighting to get the relationship between the *Enola Gay* and *their* particular narrative to be the one presented in the museum exhibit and, by extension, to be consolidated by American society. Ultimately, the display was cut off from any particular narrative except the most mundane explanation of its restoration, and the fuselage of the plane was displayed at the NASM with nothing but a video about its refurbishment.[2] As a *Baltimore Sun* reporter put it: "If you like looking at airplane parts, you'll enjoy [the new exhibit]" (Hirsch, 1995). Without connections to stable memory frameworks such as existing narratives of war heroes or of human tragedy, the exhibit was nothing but an airplane. A nuanced role for the *Enola Gay* in the collective memory of American society has failed to consolidate (Engelhardt & Linenthal, 1996; Gallagher, 2000).

Or, at least, it has failed so far. The world will not soon forget the bombings of Hiroshima and Nagasaki. Perhaps, for some societies, these events have already been consolidated into stable memory (witness Peace Park in Hiroshima), whereas for others, including the larger U.S. society, the memory of these events is still labile. But debate over this and other issues related to nuclear warfare is far from over (e.g., Huntley, 2006). The bombings reverberate through ongoing discussions of balances of power, security, nonproliferation, and disarmament; they are too heavily interrelated within existing frameworks to long remain outside the focus of consideration. There is still plenty of scope for renewed attention, and contest and negotiation, concerning the role of the *Enola Gay* specifically and the American nuclear legacy in general.

Recursion in Collective Memory Relationality

One area of social interaction that has received considerable attention from historians and other scholars as a form of collective memory is nationalism (see, e.g., the special issue of *Social Science History*, "Memory and the Nation," edited by Jeffrey K. Olick, 1998; also Anderson, 1983; Duara, 1995; Tai, 2001). They tend to understand nationalisms as quasi-fictional, socially constructed narratives—"invented traditions" in Eric Hobsbawm's words (Hobsbawm & Ranger, 1983)—or "social frameworks" in Halbwachs's terminology. Nationalisms are a form of consolidated collective memory for the political, social, cultural, and geographic entities known since the nineteenth century as "nations." A nation is a large-scale social structure that is developed over time in the material and discursive spheres. Nationalist frameworks (like other forms of generalized collective memory) recur to influence collective memory consolidation.

Consider, for example, the divided island of Cyprus. Greek Cypriots control the southern half of the island, and Turkish Cypriots control the

2. The *Enola Gay* is now permanently displayed at the Steven F. Udvar-Hazy Center, an NASM extension, near Dulles International Airport outside Washington, D.C.

Turkish Republic of Northern Cyprus. They share the world's only (currently) divided capital, Nicosia (Lefkoşa in Turkish), which sits on the United Nations demilitarized "Green Line" and which nevertheless has only one mayor (a Greek Cypriot). Although the island-wide sovereignty of the Greek Cypriot–controlled Republic of Cyprus is recognized internationally by all countries except Turkey—the Republic of Cyprus even joined the European Union in 2004—a number of groups continue to vie for the cultural and political authority to establish their version of the past and their ideal state.

There are multiple nationalisms on this island (Mavratsas, 1997). Greek Cypriots who once campaigned for *enosis* (union) with Greece were horrified at the invasion (by Turkey) and division (by the London–Zurich Agreement) of the island that their particular brand of nationalism incited; they have largely given up *enosis* in favor of a strongly Greek but independent Cyprus. Turkish Cypriots, who cannot claim a Turkish history for the island nearly as old as Hellenistic Greek history, are largely satisfied with the solution of separate autonomous states on the island because, as a numerically smaller population, they feel they would have no power under the Greek Cypriots. Middle East scholar Jim Bowman notes that "More so than Greek Cypriots, Turkish Cypriots appear more willing to let go of the past conflicts with their neighbors, making less of memory" (Bowman, 2006, p. 125). There is also a small compromise minority that takes a conciliatory middle position between these two extremes. The two main factions, however, are all-or-nothing Greek Cypriot and Turkish Cypriot nationalists, who maintain incommensurate viewpoints. As culture and tourism scholar Julie Scott (2002, p. 228) observes, "[c]asting the history of Cyprus as the story of the Greek/Turkish Cypriot national struggle presents problems of narrativity for both sides." These "problems of narrativity" extend from the particular national frameworks each side cultivates, to the potential memory items they select, and to the ways they are related to each other and back to their respective frameworks. The results are two very different consolidated memories of the same set of events.

In contrasting these two separate collective memories, political scientist Yiannis Papadakis (1994) provides an excellent example of the recurring effects of narrative on selection and relationality in his essay on the two Museums of National Struggle in the capital city. Papadakis has documented how each group's taking on the nationalism and desires of its respective "home" country is manifested in the careful selection of events (mostly raids and massacres) and their relation to specific themes (overcoming a stronger enemy, the sacredness of the dead) enshrined in each museum. The story each museum-display presents, and which visitors experience as consolidated collective memories, continue to influence political discussions. What is remarkable is how the two museums essentially place the same set of events into different sets of relations, each guided by its own, recurrent narrative. The museums are also influenced by the desires and motivations of the respective factions. Perhaps the most blatant disagreement between the two exhibits is that one museum depicts the fight as an ongoing struggle, whereas the other presents it as over.

The Greek Cypriot National Struggle Museum, which at least in the early 1990s still flew the blue and white flag of Greece, opened shortly after the island fought for and gained its independence from the United Kingdom in 1960. It was updated in the 1970s to encompass the violence that led up to and followed the Turkish invasion in 1974. The story it tells begins with the fourteenth-century BCE Mycenaean (Greek) settlement of Cyprus, and its desired ultimate end is *enosis* with Greece. But the 1974 London–Zurich Agreement not only maintained Cypriot independence (from the United Kingdom) but also formalized the division of the island into Greek Cypriot and Turkish Cypriot parts. There has been no reunion with Greece (nor even the establishment of a predominantly Greek Cyprus), and therefore the Greek Cypriot narrative attains no closure.

On the northern side of the dividing wall, meanwhile, the Turkish Cypriot National Struggle Museum, flying the red and white flag of Turkey, opened around 1980. Its narrative begins in 1571 with Ottoman conquest of the island and ends with the 1974 treaty. The moment of peripeteia for those of Turkish persuasion is 1878, when Great Britain made the island a colony and motivated the struggle for a Cypriot nation, which, according to this version of the story, was completed in 1974. "[T]he Turkish Cypriot museum's narrative closure comes to seal their history by establishing the end, independence, that is both morally good and also imputes significance to the whole of the Turkish Cypriot struggle as a 'meaningful' one in that it has accomplished its aim," concludes Papadakis (1994, p. 414).

Papadakis further points out—as we would expect considering the recursive nature of consolidation—how a coherent narrative requires the omission of counterexamples. Notably, when asked directly, individual Greek or Turkish Cypriots reveal that they have consolidated individual memories of friendly Turkish or Greek Cypriots, respectively (Bowman, 2006). In this case, potential memories of Cypriots of whatever ethnicity living peacefully as neighbors, as they did for so long, were available in the collective Cypriot buffer, but were selected for display in neither museum, thanks to the recurrent effects of social frameworks centered around hostility.

The big-picture lesson Papadakis draws from these examples is the difficulty of an entity as historically heterogeneous as a nation serving as the proper, singular subject of a collective identity. Whereas the museums claim to represent "the" history of "the" Cypriot nation (i.e., state), in fact they more accurately represent the contrasting histories of the Greek and Turkish Cypriot nations (i.e., peoples) on the island. In the context of our model, the Greek and Turkish Cypriots are two cultural–political entities sharing one island, and the curators of their respective museums constitute the most prominent members of two separate social hippocampi. Each one independently selects and relates potentially stable memory items from the buffer-full of labile ones, and each does so under the recurrent guidance of its own narrative and the animating impulses of its own entity, each complete with long-seated animosity toward the other.

The Cypriot museums case clearly shows how the same items can be consolidated differently when they are placed in different systems of relationships with each other, and it underscores the central role played by the selector/relater element of the collective consolidation process. What the Greek/Turkish Cypriot example also vividly illustrates is that museums, as memory structures created through the process of collective consolidation, store not a collection of separate objects but a set of relationships between objects that are associated to each other and to an overarching narrative, thus imbuing them with meaning. Because it is also subject to the motivations of a particular collective, the system of relationships also offers support for the claims and persistent goals of the consolidating entity. In that it presents a stable "story," distilled from a web of associations between once labile memory items, a museum is perhaps the most tangible and accessible of all collective memory structures (Kavanagh, 1996).

Conclusion

Memory consolidation on individual and collective levels involves relationality. Neither individual nor collective memory, in either labile or stable form, consists of things in themselves. Those "things"—whether words or ideas, life events or facts, museum objects or monuments—must be connected by the binding of certain features and not others. As simple as defining what something "is" appears to be, even this process is already influenced by the recursion of stable knowledge frameworks and the goals, desires, and emotions of the entity. Once memory items are sufficiently bound so as to determine what they are, they can be related to one another forming the higher-order systems of relationships that give memory its value. It is only in relation to other objects, events, or ideas that memory items contribute to knowledge, because then it becomes possible to surf between memories and to bring learned information to bear in different situations. The webs of associations the relater element establishes can be useful in themselves, and they also serve as the organized substrate from which generalized, stable memory is consolidated.

Even more than low-level bindings, systems of relationships are subject to the influences of generalized memory and of the non-mnemonic properties of the entity, whether individual or collective. The establishment of systems of relationships also may be subject to the consequences of interaction between individuals or between collectives. Whereas this effect on memory consolidation has yet to be studied for individuals, it is readily apparent on collective levels. Intergroup interaction often changes the way a collective relates items, and that in turn reshapes consolidated collective memory. Yet the influence of intergroup interaction can cut both ways: Sometimes it brings the collective memories of two groups closer together, but other times it pushes them farther apart. The larger point is that relationality is central to the consolidation process. It is the node at which all influences most effectively exert themselves.

The importance of relationality to the consolidation process is made even more apparent by considering what constitutes a useful and usable memory. An unstructured collection of memory items (e.g., representations of objects, facts, events), no matter how complete or accurate, does not constitute an efficient memory system. Indeed, it is more likely to lead to distraction or confusion than to recall of relevant information. Useful memory structures of the kind we discuss in the next chapter encode the relationships of specific items to general categories and the relationships of categories to each other. Obviously, the relationships among items must be established before they can be consolidated into a memory construct. Much of the work in forming a useful, useable, stable memory thus involves the establishment of relationships among new memory items and between new and old memory items. The relationality element is therefore central to the consolidation process.

Individual memory research tends to view this binding or relating process as taking place mainly between new items, whereas collective memory research is deeply concerned with how new items become connected to old ones. These different approaches to relationality reflect diverging disciplinary paradigms that confine memory scientists to focused experimental designs but encourage memory humanists to pursue memory contextually. As we have seen in this chapter, both individual and collective memory consolidation depend on the establishment of systems of relationships among new items and between new and old memory items. Thus, both sides of the two cultures divide (Snow, 1959) can come together around the central consolidating function of relationality.

7 Generalization and Specialization

The two core memory elements we have discussed so far, the buffer and the selector/relater, process discrete memory items such as those representing specific instances of a fact or an event. The buffer (see chapter 5) is a temporary storage device that holds and attends to verbatim, detailed, unconsolidated (new) items or already consolidated (old) items downloaded from stable memory. The relater (see chapter 6) connects items into strands and then webs of meaning. Also critical to memory is the ability to learn general rules based on sets of specific examples—in other words, the ability to derive knowledge constructs from systems of relationships among items. This is the function of the third core element in our consolidation model, which we consider in this chapter.

We call this third component the generalizer/specializer and attribute to it various properties including stable representation, generalization and loss of attribution, specialization, slow learning, fast application, and recursion. As the third and last of the three core components, the generalizer/specializer is the output stage of our memory model, and it produces the stable memories that are the product of the consolidation process. The generalizer operates by deriving a general structure, or "gist," from the items presented to it. Because generalization requires many examples, this process is slow in comparison with the processes mediated by the other components of our memory model. As it combines items into a knowledge framework, the generalizer strips them of their source or attribution. The discrete memory items then no longer require verbatim representation but have been transformed into general knowledge. The generalizer can be thought to condense or distill the knowledge contained in systems of relations that have been established between sets of discrete items. The condensed version can be more efficiently stored and accessed, and the knowledge the generalizer extracts can be applied quickly during remembering. Thus, the generalizer accomplishes the final stage of the process of memory consolidation, by which the knowledge contained in an interrelated set of items is extracted and stored for the long term in a compact, efficient, accessible, and usable form.

Conceiving of the generalizer/specializer element as the final stage of the consolidation process is useful in terms of explaining its function, but it is also grossly misleading, because memory consolidation is an ongoing process. Stating that the "end" of consolidation is a stable, generalized knowledge construct is like stating that the "end" of the thermostatic process is switching on the furnace. Just as the function of a thermostat is continuously to regulate the temperature of a house, the function of the consolidation process is continuously to reshape generalized knowledge constructs. In this more accurate view, "stable" memory is not immutable, but it does change more slowly than labile memory. The stable memory shaped by the generalizer changes slowly not only because valid generalizations take a long time to make, but also because existing stable memory can recur to influence consolidation. Through the process of recursion, existing stable memory both preserves itself and ensures its integration with new, labile memory items. Generalization and recursion are key to understanding the relatively slow dynamics of memory formation on individual and collective levels.

The generalizer possesses the complementary process of specialization. By refining categories, specialization serves as a corrective to generalization's tendency to subsume details and contexts under broader and broader headings. Based on the continuous input of new, labile memory items, specialization divides and subdivides categories so that unwieldy groupings are broken down for more precise description. Specialization is different from modulation: The former collects and merges examples of some highly defined type, whereas the latter "tags" specific, distinct instances as important. Thus, stable memory is subdivided into specialized domains over which knowledge is generalized, but within which specific, modulated items can be embedded. These properties of the generalizer account for the fact that declarative memory (knowing what; see chapter 2) can be both general and specific. With categories of various levels of abstraction and detail, in addition to modulated discrete items, stable memory both provides quick access to general knowledge and places specific instances in a broader context. Stable memory thus retains a wide array of information about the past and enables an entity to make judgments (inferences) about the present and predictions about the future.

The simplest and most intuitive examples of generalization are averages and medians, measures of central tendency that describe many specific examples with a single value. We will therefore begin our discussion of generalization and specialization in individual memory with a brief discussion of the process of deriving useful abstractions. Many disciplines in individual memory science study the process of extracting generalized knowledge: neurophysiologists have gathered evidence of generalizers in the brain; computational neuroscientists have developed models that replicate some of the findings concerning neurophysiologic generalizers; and psychologists have found generalization at work in memory and recall. We will examine broadly tuned neurons, computational perceptrons, and gist-based learning to elucidate the generalizing element's functions in the consolidation of individual memories. Generalized knowledge constructs go by many names including "classifiers," "categorizers," "schemata," "frameworks," and "narratives"; the first half of this chapter closes with a discussion of these structures and how they recur to influence the consolidation of new memories. In the second half we will explore in depth the production of stable memories in one particular collective—"science"— although our conclusions apply to other kinds of collective memory as well. Here we pay particular attention to the generalization of scientific knowledge from labile science memory and show how features of the generalization process, including the loss of attribution, the need for repetition, and the importance of generalized knowledge, are common to scientific collectives and to other social groups such as ethnic subgroups and entire nations.

A running average is a highly simplified but still useful way to describe generalization over multiple discrete items, where the specifics (such as the attributions) of each example are lost as an exemplar is created. It also nicely illustrates the idea of recursion, because the current value of a running average is a combination of the new observation and the old average. But a running average fails to capture the finer details of generalized memory and of the systems of relationships it represents, which are continually adjusted as new, labile items are introduced and must be reconciled with each other and with already existing, stable items. Because the "meaning" of an item is derived from its position in this ever-changing web of relationships, we are led to the particularly interesting postulate that stable memories can nevertheless possess changeable meanings. Finally, we will end the chapter with a discussion of narratives and how these can function as the cultural identities of societies.

Generalization and Specialization in Individual Memory

Generalization describes how individuals learn stable knowledge constructs from examples. This ability is extremely useful; in fact, it is essential to survival in the long term. Temporarily storing and relating verbatim information is valuable too, and is useful in the short term. Consider, for instance, the experience of a child who touches a red-hot heating coil (as on an electric range). The child is sure to store a representation of that specific, hot coil and of the experience of pain (using her buffer) and to establish their relationship (using her relater). These two memory components provide her the association hot coil–pain, which enables her subsequently to avoid that specific hot coil and so also the pain.

In principle, the child could learn to avoid burns by storing verbatim every instance of being burned, but there are two problems with that. First, she would accumulate, over time, a very large number of instances, and sorting through them upon encountering an object to try to match it against each of her verbatim records of every object that ever burned her is a very inefficient way to avoid a repeat experience. Second, unless she encountered the same object she had experienced before, she would not find a match and would be obliged to burn herself again in order to gain the association of this new object and pain. Fortunately, because in addition to a buffer and a relater she has a generalizer, she can abstract general principles from her experiences, and she soon learns that anything glowing red is likely to be hot and to cause pain if touched. Rather than having laboriously to sort through a tremendous catalogue of specific instances, she can quickly and efficiently apply her general knowledge. Getting smart is the best way to avoid getting burned!

Although generalizing over many separate experiences is an efficient way to store and access stable memory, specific items and episodes can also be represented in an ungeneralized form in stable memory. Put another way, loss of attribution for some items does not imply loss of attribution for all items. An adult, for example, may never forget the specific episode when, as a child, she touched a red-hot range-top burner. As we mentioned in the last chapter, certain items and episodes can be modulated by emotion and/or attention and become consolidated as discrete (distinct) memory entries, which are interrelated with more generalized knowledge in stable memory. The hot-coil example fits nicely with the viewpoint of psychologists Hellen Williams and Martin Conway (2009), who consider autobiographical memory to be composed of episodic memory (specific episodes) and autobiographical knowledge (generic, schematic knowledge of personal history). Autobiographical memory may be the richest and most complex kind of individual, generalized memory. Categorization, which we consider next, may be the simplest.

In its most basic form, the generalizer component can be considered as an input–output device: You provide a specific input, and it gives you back a generalized output. Although the generalizer is able to learn many different kinds of input–output transformations, it is easiest to think of it as a categorizer: You provide a specific input, and it gives you back the category to which that input most likely belongs. Imagine, for example, a visual sensory perception of a dachshund. Some of the features of the perception are unique to the breed of dachshund (characteristic sausage shape), but some are true of all dogs (fur, four legs, wagging tail). The verbatim, episodic memory mediated by the buffer and relater components can temporarily store a representation of "that dachshund" and bind it to associated perceptions ("the person walking that dachshund"). But a different system is required to learn higher-level abstractions such as the idea (or class) of "dog" or "mammal." Recall that the inability to create such categories was Funes's cognitive handicap (Borges, 1942/1962).

Versions of generalizer/specializers appear in neurophysiology, computational neuroscience, and cognitive science. Neurophysiologists find evidence of generalization at the level of single neurons. The activity of certain neurons in the temporal lobe of macaque monkeys, for example, greatly increases when the monkey is looking at a face. The species of the face is not crucial: The cells will fire in response to macaque, human, or even smiley faces (Tanaka et al., 1991). You might say that the firing rate is proportional to the probability that the perceived item matches the stored category of "face." Thus, the cells will also fire (albeit more weakly) in response to profiles or upside-down faces. They will not, however, fire in response to lines or hands or isolated noses. These temporal lobe neurons receive inputs from lower-level visual cortex, which extracts and combines visual features such as colors, edges, boundaries, and curves. By generalizing over such input combinations, the "face cells" can abstract the category "face" based on features common to previously observed faces. The attributes that distinguish specific faces have been lost, and "face" has been elevated to a stable classification, without retaining the unique, individual input features.

Some computational neural network models exhibit generalized feature-learning (for background, see Anastasio, 2010). For instance, a two-layered network can learn to categorize if each neural element (unit) in the output layer becomes a specialist for its own, distinct subset of a potentially large number of input patterns (e.g., Rumelhart & Zipser, 1985; Kohonen, 1988). The same output unit could be activated by input patterns corresponding to images of either a poodle or a Pomeranian, because both share characteristics in common (i.e., fur, four legs, barks, likes bones, etc.). In other words, this two-layered network can abstract a general category from its members' shared characteristics. Such a network can be thought of as a categorizer because a single output unit represents a large number of input patterns that share common features.

To learn to categorize inputs based only on the overlap of common features, two-layered networks can be trained using unsupervised algorithms, which require only the presentation of input patterns; the network can automatically come up with its own classifications. If the categorization requires the interaction of elements of the input patterns, then supervised algorithms are needed to train it. Supervised algorithms are much more powerful than unsupervised algorithms, but the former require input–output pairs for training (Rumelhart & McClelland, 1986). In our model, such pairs of properly associated patterns are supplied by the relater component. Given such pairs, and if the interactions between pattern elements are not too complicated, then a two-layered network can be trained to categorize them using a simple supervised algorithm known as the delta rule. If the interactions required for classification are more complicated, then the network needs to have three or more layers of units and must be trained using a more sophisticated algorithm known as back-propagation.

A multilayered neural network trained using back-propagation is commonly known as a multilayered perceptron (see also chapters 2 and 4). Such an adaptive neural network, implemented on a computer, is capable of learning some truly amazing classifications, such as automatically "reading" zip codes by classifying as an Arabic numeral each handwritten character at the bottom of a letter's address (LeCun et al., 1989). Learning to categorize is handy for technological devices, but it is crucial for the survival of organisms. For the zip-code reader, engineers supply the input–output pairs. For a real brain they would be supplied by the relater element (presumably the hippocampus), perhaps with a little help from its friends, and especially from parents. For example, a parent points out specific dogs and repeats the word "dog," providing input–output pairs that the child's brain uses to form the category *dog*. After many presentations, the child learns the association and, upon seeing any dog, instantly shouts, "dog!" to the delight of her parents!

As we explained previously (especially in chapter 3), much learning is socially mediated: parents aid the relational processes occurring in the hippocampi of their children; teachers help their students; and members of social groups assist each other. But in the end, the learner's own hippocampus must represent the relationships in order for them eventually to become generalized. The associations are not one-to-one but many-to-one: the child could represent and distinguish many different types of dogs and still assign them the same category name. Thus, generalization can be simulated as pattern association using a multilayered perceptron, where the associations between the patterns are established through relationality. This is another example of how interdependent the consolidation elements are on one another.

A multilayered perceptron, trained using a supervised learning algorithm, learns only those categories that the output patterns specify. However, the network is free to create "internal representations," which are patterns of active units located in its internal (neither input nor output) layers (also known as hidden layers). These "hidden units" can become specialized for subcategories of input patterns and can mediate specialization in the network. David Rumelhart and P.M. Todd (1993) describe one such example. By generalizing over their common features, a multilayered perceptron trained with certain discrete input patterns (trout, bass, oak, maple, robin, blue jay, daisy, rose) learned to categorize them correctly (fish, trees, birds, flowers). It accomplished this learning by forming internal representations for attributes such as "swims," "flys," "has roots," and so forth. However, when the network was given an odd input—in this case, a penguin—that simultaneously belonged to two of its categories, it was initially thrown into disarray. Fish or bird? The system spit out contradictory outputs in a sort of "contest and negotiation" between the two existing categories. Given enough training time, however, the perceptron was able to create a third, specialized internal representation for "a bird that swims," which simultaneously activated both the "fish" and "bird" categories as required for this specific input. Owing to the correspondence between the behaviors of these artificial neural networks and children learning categories, Rumelhart and Todd (1993) suggested that something like the specialization that occurs in multilayer perceptrons also occurs in the actual nervous system. The organization and reorganization required for generalization/specialization in artificial neural networks is presumably what makes it such a slow process in real brains as well.

The cognitive science literature also addresses this generalizing feature of memory. Human ecologists Charles Brainerd and Valerie Reyna have developed a "fuzzy-trace theory" of memory accuracy (Howe et al., 1992; Brainerd & Dempster, 1995; Brainerd & Reyna, 2001, 2005). Their theory distinguishes between verbatim memory—what we call buffered storage—that contains all the details, and the "gist-based" generalizer, which contains only general information. According to Brainerd and Reyna, parallel verbatim and gist representations of the world both form in response to stimuli, and during recall both of these representations are used to guide behavior. Over time, however, the verbatim trace degrades, whereas the gist trace is better preserved. This is the cognitive version of the trace theory discussed in chapter 6 for the relations between the medial temporal lobe (the details) and the neocortex (generalizations) (Alvarez & Squire, 1994). A category could be thought of as a "gist." For instance, in trying to explain a wolf to a friend who has never seen one, you might say "it is like a dog" and your friend could get the "gist" of your description through invocation of the category. But the "gist" is better thought of as "the skinny," or as the essence of a situation. This brings us to schemata.

Bartlett's Legacy: Schemas

Along with classifiers, knowledge frameworks, paradigms, and narratives, schemata (or schemas) are a type of generalized memory. A schema can be thought of as a "general plan" for a whole class of items, and an individual has many of them (cf. Brewer & Nakamura, 1984). You have, for example, a "chair schema," which represents the general plan for a chair: four legs, a seat, a back, maybe some arms, and so forth. Your chair schema is emphatically *not* a detailed image of any particular chair from your experience. Rather, it is a Platonic ideal abstracted from all the chairs you have ever experienced. Neither is it the average of all the chairs of your experience. It is smarter than that: It is a framework extracted from the system of relationships among the parts of a chair, and of a chair in its proper environment, that constitute what it means for an object to be a chair.

Perhaps the most intriguing and most elusive schema is the "self-schema," which can be thought of as a framework that defines the identity of an individual. As a generalized memory, a schema is created by the consolidation process; and again as a generalized memory, schemata can recur to act on the buffer and relater and influence memory consolidation. Schemata have revealed themselves to psychologists first through their formation as generalized or abstracted knowledge constructs and later through their recursive influence. Though modern analyses of the generalizing element have provided detailed insight into schemas, the basic concept is much older: It is a central feature in Frederic Bartlett's 1932 book *Remembering*, in which he deduces the existence of schemas and suggests how they function in recall. Working from this foundation, we can argue for the influence of schemas on consolidation.

Bartlett was unsatisfied with his predecessor Hermann Ebbinghaus's methods of memory investigation through list learning (see chapter 2). This paradigm—still in use—attempts to characterize memory objectively, but Bartlett considered the interesting aspect of memory to be its subjective nature. He characterized Ebbinghaus's work as a "study of the establishment and maintenance of repetition habits" (Bartlett, 1932, p. 4) rather than as a study of memory per se. In his own work, Bartlett used what he considered a more naturalistic approach, presenting subjects with complex pictures or stories and measuring their ability to recall them accurately. In one experiment, he showed a series of photographs of military officers' and enlisted men's faces to a set of subjects. Over the course of a month, Bartlett re-interviewed the subjects and had them describe the faces they had seen. As time passed, their descriptions became less and less detailed and more and more centered on a single evocative feature or supposed personality trait (e.g., "the moustache," "good-humoured"; p. 53-57). The complex figures were reduced over time to only the features necessary to distinguish one face from another—not verbatim sets of traits but "sketches" that abstracted the defining features.

In an effort to explain these behaviors, Bartlett developed a theoretical construct he called a "schema." A schema is "an active organization of past reactions . . . always supposed to be operating" (Bartlett, 1932, p. 201). This description immediately evokes Thomas Kuhn's conception of paradigms in scientific communities (Kuhn, 1962/1996; see also chapter 4). Schemas provide an individual not with specific images of past events but with abstracted, generalized knowledge garnered from many past events. New information is not stored verbatim in a schema but instead interacts with existing schemas to produce a mental representation that contains features from both the old and the new experiences (Head, 1920).

The influence of schemata is nicely illustrated in an experiment that follows the Deese–Roediger–McDermott (DRM) paradigm. In the original DRM experiment, a lure word that is related to other words on a list is incorrectly remembered as also being on the original list (see chapter 6). In a visual rather than verbal example of the DRM paradigm, Brewer and Treyens (1981) found that subjects "remembered" that there were books on the shelves in an office they had visited, when in fact the shelves were filled with various objects but no books! The authors explained this finding by hypothesizing an "office schema," or a mental model of a typical office, in which books are found on shelves. The subjects drew from their "office schema" in "remembering" that there were books when in fact there were none.

Bartlett's schema theory represents the past as an "organized mass" (Bartlett, 1932, p. 201). Moreover, it is a flexible mass (or construct) that changes over time and with new input—like a running average, but smarter—and there are many such malleable schemata, representing everything from sofas to the self. Perhaps Bartlett's greatest contribution, the idea of schema has profoundly influenced thinking about memory. To complement his notion of generalized schema, Bartlett also provided the initial psychological insights into generalization as a process.

Bartlett's Schemata in Action

Bartlett was the first to describe the forward-going effects of generalization and the recursive effects of already generalized memory on what we now understand as the consolidation process. As the active-verb title of his book demonstrates, he was interested in characterizing the reconstructive thought process of subjects during the recall of information. He proposed that memory formation involves the creation (or modification) of generalized schemas and that new experience reactivates schemas that can then be used generatively—not just to reconstruct representations of previous experience but to create new "riffs" in labile memory based on current input. Bartlett called this "turning round one's own schema" (Bartlett, 1932, p. 202), and the idea presaged recent findings that subjects can use schemata to "imagine" generatively the outcome of events by mentally simulating them (Brewer & Pani, 1983; Brewer & Nakamura, 1984; Schacter et al., 2007). Bartlett's famous "repeated reproductions" studies show— mainly by analyzing inaccuracies in recall—how the generalizer streamlines memory content and how the recursion of stable knowledge frameworks affects memory formation and remembering.

The method of repeated reproduction is simple: A subject is given a picture to look at or short story to read and is then asked to reproduce the image or story as accurately as possible over a series of time intervals. Bartlett found that, with each reproduction, the content of the original story changes in somewhat predictable ways. Here is a sample passage:

One objection to the views of those who, like Mr Gulick, believe isolation itself to be a cause of modification of species deserves attention, namely, the entire absence of change where, if this were a *vera causa*, we should expect to find it. In Ireland we have an excellent test case, for we know that it has been separated from Britain since the end of the glacial epoch, certainly many thousand years. Yet hardly one of its mammals, reptiles or land molluscs, has undergone the slightest change, even though there is certainly a distinct difference of environment, both inorganic and organic. That changes have not occurred through natural selection is perhaps due to the less severe struggle for existence owing to the smaller number of competing species; but if isolation itself were an effective cause, acting continuously and cumulatively, it is incredible that a decided change should not have been produced in thousands of years. That no such change has occurred in this and many other cases of isolation seems to prove that it is not itself a cause of modification. (Bartlett, 1932, p. 166)

The debate alluded to here is of course part of the fallout from the publication of *On the Origin of Species* (1859) by English naturalist Charles Darwin (1809–1882) some 60 years earlier. The quotation comes from British evolutionary biologist Alfred Russel Wallace (1889, pp. 151–152). Co-founder and later recanter of evolution by "natural selection"—the term he encouraged Darwin to use—Wallace (1823–1913) later refused to believe that geographic isolation alone can account for variations (it can). "Mr Gulick" refers to John Thomas Gulick (1832–1923), a missionary to Japan and probably the most famous American evolutionary biologist you have never heard of (Romanes, 1892–1897; Gulick, 1905; Hall, 2006a, 2006b). One of Bartlett's subjects rendered the passage after only three subsequent reproductions this way:

Mr. Garlick [sic] says isolation is the cause of modification of species. This seems proved by the test-case of Ireland with regard to snakes toads and reptiles [sic]. (Bartlett, 1932, p. 167)

The most obvious difference between the original and the remembered passages is that the latter is much shorter. It represents the "gist" of the passage with most of its details omitted. The process of generalization has abstracted the meaning from the original item and discarded the "rest." Bartlett consistently found that the remembered (i.e., consolidated) item was an abbreviation or summary of the original item. His findings provided some of the earliest well-documented support for the idea that generalization is a major component of the memory formation process. They also provided some of the earliest evidence of memory inaccuracies.

The next most obvious difference between the original and remembered passages is that the latter is wrong. Not only are details recalled inaccurately, but the main thrust of the passage is recalled backward. The subject substituted the opponent for the original author and thereby turned the argument of the passage completely around! Bartlett found many such egregious errors in recall. He called them "radical alterations" and described them this way:

Epithets are changed into their opposites; incidents and events are transposed; names and numbers rarely survive intact for more than a few reproductions. . . . At the same time, the subjects may be very well satisfied with their efforts, believing themselves to have passed on all important features with little or no change, and merely, perhaps to have omitted unessential matters. (Bartlett, 1932, p. 175)

Bartlett explained many of the recall errors he observed by what he referred to as an "effort after meaning" (Bartlett, 1932, p. 20). According to Bartlett, a subject takes whatever he can remember, whether accurately or inaccurately, and makes an "effort after meaning"—that is, he attempts to construct from it a reproduction coherent to him. (The subject quoted above obviously knew very little of the actual debate.) We attribute this "effort after meaning" to the non-mnemonic cognitive processes entities use to make sense of the world (for more detail, see chapter 8). Because subjects believe themselves to be careful and correct in their reproductions, they tend not to check their remembering. Thus, errors become fodder for future consolidation, and because of the recursive nature of remembering and consolidation, such mistakes are compounded with each new iteration. Although recursion serves to preserve stable memory and integrate it with labile memory, Bartlett's example amusingly shows that the recursion of already consolidated schemata can also lead to memory inaccuracies.

The subject rather delightfully remembers "Mr Gulick" as "Mr Garlick." He obviously would not have made this mistake if "garlic" was not already part of his consolidated memory. The subject also remembers "snakes toads and reptiles," when the organisms mentioned in the original passage were mammals, reptiles, and land mollusks. Clearly, the subject has an (apparently not very good) "animal kingdom" schema and draws from it, as in the DRM paradigm (see chapter 5), to add animals not present in the original passage, even as he subtracts other animals that were present. (We will leave aside his error on the presence of snakes in Ireland.) This example, along with many others from Bartlett's catalogue, demonstrates the process of formation of new generalized memory and the effects of recursion of already formed generalized memory.

The work of Frederic Bartlett, done in the early part of the twentieth century, elucidates clearly the role of recursion in the process of memory consolidation. This is due to Bartlett's willingness to assess the interests, biases, and tendencies of his subjects qualitatively—for that was the only way they could be assessed. Bartlett was also trained in anthropology, and he noted with great interest how the ethnic backgrounds of his subjects could distinctively affect their recall. We contend that memory consolidation cannot be fully understood without understanding the recursive influence of stable memory constructs such as self-schema (identity), narratives, and social frameworks. It should be possible, using modern techniques, to assess more quantitatively the "identities" or "frameworks" of subjects and the influences these exert on their recall, but we certainly do not mean to suggest that this would be straightforward. Given the recursion that we propose, we expect self-remembering to be complex indeed.

Real and Imagined Narratives as Generalized Knowledge Constructs

One common type of construct consolidated in the generalizer is that of "narrative," which includes internally coherent descriptions of situations, happenings, and courses of events. The field of literary criticism defines narratives rigorously as stories that take place in time with beginnings, middles, ends, characters, and so on (Forster, 1927). In narrative psychology—one scientific subfield that has consistently emphasized the recurrent influence of these stable knowledge constructs—it is enough for a narrative "to explain the way things are." Championed by Ted Sarbin (1911–2005) (Sarbin, 1986) and Jerome S. Bruner (Bruner, 1986, 1990), narrative psychology proposes an approach to the psychology of individual memory and identity that appreciates the sense-making properties of narrative. It provides a useful descriptive framework for understanding human memory, but its explanatory value has been largely unrealized.

The main conclusion of much of Sarbin's work is that many of the normal functions of human remembering closely resemble the same processes used in the creation of a personal, fictional narrative. (Hayden White has made the same argument about history; see H. V. White, 1973, 1980, 1987.) For Sarbin, the verisimilitude for a subject of an imagined or a remembered event is determined by how thoroughly and personally the subject believes in the accuracy of the story. In 1998, he co-edited with Joseph de Rivera a compilation of articles entitled *Believed-in Imaginings: The Narrative Construction of Reality* (de Rivera & Sarbin, 1998), in which each article drew parallels between the act of remembering and the act of imagining. In his chapter, de Rivera cites a study (Rozin et al., 1986) in which participants were asked to pour sugar into two identical containers and then choose one to label "sugar" and one to label "cyanide." This gave the subjects two possible explanations of the visual reality in front of them: the memory that both vessels held sugar, and the imagination that one of the vessels held poison. When the subjects were asked to sample the substance in each vessel, they were more hesitant to taste the sugar labeled "cyanide" despite having a factual recollection of filling and labeling the two identical containers themselves!

To narrative psychologists, this experiment demonstrates that remembering and imagination cannot be fundamentally different processes. If they were, test subjects could easily apply a "true" value to the memory and a "false" value to the imagination, and they would not hesitate to sample the substance they had labeled "cyanide" just minutes before. We have already explained that remembering is best understood as a constructive act, and one that is prone to inaccuracies and biases (as in Bartlett's work). Sarbin and other narrative psychologists conclude that imagination, like memory, draws on previously learned models of the world. To explain this in the context of our model, the entity would draw on already consolidated, generalized knowledge constructs (categorizers, frameworks, schemata, narratives, etc.) both to remember and to imagine. As Sarbin and others have shown, the memory process is not privileged, and if an imaginary label can override a true memory, then the distinction between remembering and imagination becomes very murky indeed.

Proof-positive that "real" and "imagined" memories do in fact consolidate similarly—that there are not separate processes for the consolidation of fictive and nonfictive information—is found in "false memory" research. In the late 1980s and early 1990s, a series of criminal proceedings were initiated on the testimony of victims who claimed to have recovered long-repressed memories of childhood abuse. Some psychologists questioned whether such repression was possible, and they suggested that the memories of childhood abuse were not recovered but instead created in adulthood, often due to the intervention of some form of psychotherapy (Pendergrast, 1996; de Rivera, 1998). Psychologists had already established that it was possible to generate minor false memories, such as identifying a word as previously included in a list when it had not been (i.e., DRM; Deese, 1959; Roediger & McDermott, 1995). However, it was not until the 1990s that psychology and law expert Elizabeth Loftus demonstrated with her landmark studies that vivid and complex false memories could indeed be created (Loftus, 1996a, 1996b; Loftus & Pickrell, 1995; Loftus & Ketcham, 1994).

In Loftus's classic experiment, experimenters read to subjects four scenarios involving vivid childhood memories. Three of the episodes were generated through interviews with the test-subjects' parents, but one of the episodes was entirely fictional. Subjects were later asked to recount the episodes, and 25% of test subjects claimed to have real memories of false events. Some even provided details not present in the original fictive episode as they used their autobiographical narratives or identity self-schemata to fill in details that made the "memory" as a whole more consistent and lively. Rather than reject the fictitious items as such, some subjects incorporated them into their personal identity narratives. This research shows that an individual's autobiographical narrative, like an autobiography, is to some extent constructed.

The analysis of autobiography in literary criticism is probably the most well-developed area of individual memory research in the humanities. Much recent work focuses on notions of identity and self. For instance, James Olney, a scholar of English, French, and Italian literature, relies on the "life-writings" of three key authors to trace the development of autobiography as a self-conscious genre in his book *Memory & Narrative* (1998); along the way, he demonstrates how remembering enables the production of narratives that recursively shape memory. First, in his *Confessions* (397–398), post-Nicene church father St. Augustine of Hippo (354–430) muses on the ability of his memory to allow him to remember the events of his life, which he then narrativizes. Second, French Enlightenment philosopher Jean Jacques Rousseau (1712–1778) in his *Confessions* (1769/1782), *Dialogues* (1782), and *Reveries* (1782) redirected autobiography toward a more intensely personal, self-revelatory literature. Third, Irish author Samuel Beckett (1906–1989) in *The Company* (1979), among other writings, focuses on the self-as-project; on individual identity as never fixed but always being re-created (rather as Bartlett described schemata).

All three men not only animate their contemporary identities with narratives from and about their pasts, but they also address the recursive nature of remembering and narrating. As autobiographers, they drew on their consolidated individual memories to recollect that which they crafted into their self-narratives. These autobiographies, as newly consolidated memories, subsequently influenced how they saw and re-remembered themselves both as individuals and as authors. Other work in the humanities stresses the social influences on autobiography. For example, English scholar Paul John Eakin reminds us in *How Our Lives Become Stories* (1999) that every self-narrative references other selves and other narratives: The writing of an autobiography is no more an act isolated from society and social influences than are individual memory consolidation and remembering. For the sake of our analogy argument, we will not discuss the interaction between individual and collective memory in detail, but we return to the idea of narratives in collective memory at the end of this chapter.

Scientific work on the nature of the generalizer element of individual memory thus spans the range from neurophysiologic studies showing the category-specific activity of single neurons to studies in psychology and cognitive science indicating that stable memory is both flexible and complex, representing the relationships among categories and selected, specific items in behaviorally useful, readily applicable ways. The formation of efficient and effective generalizations requires the continuous rehearsal of associations among memory items. In our framework, the relationships needed to train the generalizer are established by the relater, using new and old memory items held in a labile state in the buffer, and continually presented to the generalizer. The formation and further development of stable, usable, and generalized knowledge constructs and their associated stable memory structures is the object of the consolidation process. There is evidence for this process among collectives as well as in individuals.

Generalization in Collective Memory

When confronted with the term "collective memory" for the first time, one could justifiably conclude that it means "the memory of a collective" and could, with equal justification, equate that with "the history" of a collective. But equating "collective memory" and "history" runs counter to a distinction many historians make, in which "history" is a carefully researched, comprehensive, and accurate account of the past whereas "collective memory" is an oversimplified, inaccurate version of the past that serves the identity claims of a group such as a people or a nation (e.g., Novick, 1999; Wertsch, 2009). We will consider this particular debate more extensively in chapter 12. Here we emphasize that consideration of the process of memory formation, rather than of memory itself, leads us to see many different kinds of social constructions as falling into the category of collective memory, including history.

Any process a collective undertakes that begins with a set of items, continues with the establishment of relationships among those items, and ends with a condensed and coherent abstraction from those relationships is a process of collective memory formation. In this light, history is clearly a form of collective memory, because its production begins with a set of memory items, such as those made available in libraries and archives; continues with the establishment of relationships among those items, by historians working as a "social hippocampus"; and ends with the construction of a narrative, presented in a memory structure such as a book or documentary. A distinction should be made between "analytic history," which is produced by historians after thorough research, and "popular history," which arises through the interactions of the society more broadly, but in both cases a set of facts is put together and a coherent story emerges. The "analytic history" may be better researched, and therefore more complete and accurate, than the "popular history," but both were formed according to a process of collective memory consolidation so both would be considered as forms of collective memory.

Despite being thoroughly researched, an analytic history is a generalization, because it is not merely a collection of unedited facts and pieces of evidence (Bloch, 1953/1964). An analytic history begins with research and by assembling the facts, but historians then select and relate pieces of evidence together to construct a coherent, narrativized version of past events. A historian of the Civil War, for example, does not list every letter from every soldier that ever found its way into an archive. Instead, the historian develops a general picture of the past based on the relationships he establishes between many different pieces of evidence. This is a consolidation process, and it is collective consolidation, because many individuals were involved in creating the evidence in the first place, in preserving it in archives, and in instructing and interacting with the actual author of the history who, in any case, usually writes about events he was not alive to witness. Indeed, a historian uses not only first-hand accounts but also considers syntheses of original material by previous historians in forming his own synthesis. The larger point is that a great many details are lost in the process of constructing a coherent narrative, so the end result of the process of producing a history is a generalization.

Although many details are lost in writing a history, some are retained, just as the process of modulation tags certain items for consolidation in individual memory as distinct items, which retain their unique characteristics but are nevertheless situated within a generalized framework. James Wertsch (2009), who uses the term "collective memory" the way we would use the term "popular history," sees it as being composed of "specific" and "deep" components, where the former is a distinct memory whereas the latter is a "generalized gist" or a "schematic narrative template." A specific collective memory might be that Joseph Stalin led the Soviet Union during the Great Patriotic War (against the invading Axis states during the Second World War). As an example of a deep collective memory, Wertsch offers the "expulsion of foreign enemies" template, which he sees as a major theme in Russian collective remembering that developed as the result of generalization over many instances where the Russian people had to drive off invaders. The specific/deep division Wertsch proposes is consistent with our view that collective (and individual) declarative memory is composed of discrete items embedded within a generalized framework. He suggests that the specific/deep distinction is a characteristic of popular history, but it must operate similarly in analytic history, as historians cite specific events as exemplars of more general trends and identify specific personages as having had significant influence on entire populations.

The distinction between analytic and popular history becomes important when considering, among other issues (see chapter 12), their relative level of detail. An analytic history is, as it must be, a generalization from, rather than an exhaustive listing of, all available facts. Still it is much more detailed than a popular history. Moreover, an analytic history can serve as a corrective to a popular history, and historians make an essential contribution by providing additional details that clarify the view and deepen the understanding that regular folks have about their shared pasts. Considering differences in thoroughness, accuracy, and detail, it is reasonable to suggest that a popular history is more generalized than an analytic history, although the formation of both, as types of collective memory, involves generalization.

Generalization and Modulation in Collective Memory

The comparison of the consolidation process with a running average is useful for emphasizing the loss of attribution that occurs when labile, discrete memory items are generalized over, in order to form stable memory (knowledge) constructs. But, as pointed out in previous sections of this chapter, loss of attribution for some items does not imply loss of attribution for all items. Most items lose their attribution, but some especially noteworthy items are preserved more or less intact and remain as distinct items in stable collective (or individual) memory. They can endure continual consolidation over long periods, even withstanding changes in the system of relationships of which they are a part, but shifting relationships imply changes in the meanings attached to distinct, stable memory items. Three examples from American collective memory (specifically popular history) make this clearer.

Historian Barry Schwartz has chronicled the changeability of the meaning of a particular, stable American collective memory item, Abraham Lincoln (1998, 2000, 2003, 2008). Lincoln's presence on the penny, the five-dollar bill, and the Mall in Washington, D.C.—not to mention in countless classrooms, museums, books, and school and town names—ensures that the sixteenth president of the United States will not be forgotten anytime soon. The more interesting question is how exactly he has been collectively remembered: In the Progressive Era, for instance, Lincoln was an Everyman success story, whereas to civil rights leaders, Lincoln was the Great Emancipator (Schwartz, 2000). Such shifting meanings are not a matter of "anything goes." As Schwartz has pointed out, the life Lincoln lived has constrained the possibilities for remembering him. But there are sufficiently many possibilities that at different times over the decades since Lincoln's death, various social groups have emphasized different details of his memory to suit their particular contexts. "Lincoln" is a stable figure; his meaning is changeable.

Few historical figures arouse as much controversy in contemporary America as the Italian (in his time, Genoan) explorer Christopher Columbus. Owing, perhaps, to the American tendency to give credit where it is due, most Americans still admire Columbus for "discovering" America (after the Native Americans and the Vikings, of course), but judgment on the character of this iconic figure depends, in part, on the ethnic group to which a particular American belongs. It also depends on age group: Whereas some older Americans still admire Columbus as an explorer, many younger Americans consider him as the granddaddy of all exploitative European colonizers of the New World (e.g., Schuman et al., 2005). This generational shift must be seen against a broader historic background. John Noble Wilford (1991) has documented the waxing and waning popularity of Christopher Columbus over the centuries.

Feminist and cultural studies scholar Ella Shohat has made a similar observation concerning cinematic portrayals of Cleopatra (69 BCE to 30 BCE): "Each age, one might say, has its own Cleopatra, to the point that one can study the thoughts and discourses of an epoch through its Cleopatra fantasies" (Shohat, 2006, p. 167). So, for instance, the actresses playing Cleopatra in films about the ancient Egyptian queen were white until the 1970s, when Black Power and Afrocentricism asserted that black could be strong and beautiful. Various groups have manipulated Western collective memories of Cleopatra as they seek to consolidate a particular historical meaning for her life. More recently, in her book *Cleopatra: A Life*, biographer and historian Stacy Schiff (2010) emphasizes the Egyptian queen's political acumen and her Macedonian Greek heritage. Is yet another cinematic portrayal of Cleopatra in the offing?

From the perspective of our model, Abraham Lincoln, Christopher Columbus, and Cleopatra VII Thea Philopator are modulated memory items. They have all made such an impact on events that their presence as discrete, distinct items in stable collective memory is all but ensured. What can change, though, are the relationships between Abe, Chris, and Cleo and other new and old memory items, whether discrete or generalized. These systems of relationships change over time, and so do the social frameworks consolidated from them, just as Bartlett postulated that individual schemata change over time. The changeability of meaning of collective memory items, as well as other phenomena such as generalization, specialization, modulation, and recursion that apply to collective memory more broadly, are illustrated by the example of the consolidation of scientific collective memory. We revisit this topic in the next section.

Consolidating Stable Collective Science Memory: Take 2

Knowledge, or science, is no exception. To be sure it is not to be confused with its history. But it is not true that the scholar or the scientist operates only from the perspective of the present. Knowledge is too collective a project not to give the scholar, even when he concentrates on a new experience or on original meditations, the feeling of following directions of research and of continuing a theoretical effort of which the origin and point of departure are located previous to himself. Great scholars place their discoveries in the chronology of the history of knowledge. In their eyes, scientific laws represent not only elements of an immense structure situated outside of time; they also perceive behind these laws and along with them the entire history of the efforts of the human spirit in this domain. (Halbwachs & Coser, 1925/1992, p. 176)

In the context of trying to understand collective consolidation as a process, perhaps one of the most useful case studies concerns the uptake of new knowledge into science textbooks. Because scientific knowledge can be relatively well described, its development provides an excellent example of collective memory formation, as both Maurice Halbwachs (above) and Thomas Kuhn (1962/1996) suggest (see chapter 4). Kuhn focused on "revolutions" in science, but his framework also applies to the more gradual ways in which scientists assess new data and re-assess old ideas as they continually develop their shared understanding of the natural world. This "understanding" is emphatically *not* the entire catalogue of findings but a synthesis of facts and ideas into an efficient and usable generalization—a paradigm—that is packaged into textbooks, taught to students, and carried around in the brains of scientists themselves.

The development of scientific knowledge also provides an appropriate example of generalization in collective memory formation, because the tendency to generalize is so strong in science. As discussed above, an analytic history is also a generalization, but its level of generalization is more limited. Although sweeping generalizations from history occasionally appear (a case in point is *The Rise and Fall of the Great Powers* by Paul Kennedy, 1987), historians and other humanists more often prefer to complexify and strive to enlighten by adding facts and details that explain specifically "how things really were." In contrast, scientists' animating impulse is to find general principles that can account for broad ranges of observations on the natural world. These principles are abstracted from data by scientists who select and relate them (with some amount of contest and negotiation) and ultimately write them into science textbooks (printed or, increasingly, digital). The effects of scientific memory consolidation can be discerned in the changes made in textbook coverage in various scientific subdisciplines.

Early textbook studies, such as Monica Winstanley's (1976) on the entrance of James Watson and Francis Crick's DNA double-helix hypothesis (1953) into British scientific publications, used citation analysis to count and graph the appearance of this new knowledge. Winstanley found that, already in 1955, almost every research-level biology textbook included a reference to these two molecular biologists and to DNA as the genetic basis of life, whereas undergraduate and primary school texts picked up this information gradually over the next decade. Meanwhile, popular science literature, reflecting its heterogeneous quality, mentioned Watson (1928–), Crick (1916–2004), and genes "sporadically." No serious biology text would be considered complete today without a description of genes, DNA, and the Watson–Crick double helix. This (at one time) new information was selected and related to (then) established knowledge and has been consolidated with it as part of stable scientific memory.

Collective scientific memory (labile and stable) is the synergistic combination of all the scientific topics currently in debate and discussion, whether at conferences, in journals, or in textbooks. Just as with individual memory, the consolidation of stable scientific memory requires processes (and structures) including buffered storage and attention (in journal articles), selection and relation (at meetings, conferences, and in review essays), and, finally, generalization and specialization (in textbooks). The content of science textbooks, taken together, can be readily described as generalized knowledge and specific knowledge. The constructive tension between generalization and specialization in science is apparent from the division of natural science into ever more specialized subdisciplines, each with their own scientific paradigms and textbooks. Within each, certain core concepts are distilled from the vast "mash" of observations and ideas brought together by the community of scientists. This generalization process necessitates loss of most of the details that accompanied the original experiments, observations, and ideas.

How did the DNA double helix come to be consolidated as a generalized concept in collective scientific memory? It was through the efforts of an entire community of scientists. As was true for the multilayer perceptron learning the category "face," all those molecular geneticists do not receive credit in the texts for their work: Consolidation of collective science memory entails—even requires—a loss of attribution for something like the protein-encoding properties of DNA to become established as a fact. French sociologist of science Bruno Latour addressed loss of attribution in the creation of scientific facts in *Science in Action* (1987) and, with Steve Woolgar, in *Laboratory Life* (1979) (see also Fleck, 1935/1979). A "fact" is a statement needing or allowing no further modification, such as who discovered it, when, where, and under what conditions. For instance, whether or not we append Isaac Newton's name to the "law of gravity," we still believe it acts on us and every other material object. Gravity is an accepted fact, and we trust it is true with or without attribution.

Latour and Woolgar use as an example the structure and function of what is now called thyrotropin-releasing hormone (TRH). As long as neuroendocrinologists Richard Guillemin and Andrew W. Schally waged a war in the 1960s for priority in describing the "true" characteristics of this neurohormone—which Guillemin called thyrotropic-releasing factor (TRF)—statements had to be qualified: "Schally believes . . ." or "In his most recent article, Guillemin claims . . . ." TRH is considered a fact now that unmodified statements such as "TRH is a tripeptide with the structure glutamine-histidine-proline" can be made. TRH has had its existence and physiologic relationships so well established that it has become a stable concept, a fact that has been generalized from many experimental results. TRH has largely lost the attribution to its original discoverers and is now generally "true."

Although loss of attribution is the rule in both generalization and specialization, some collective science memories are stored as distinct name–discovery pairs. In molecular genetics, "Watson and Crick–DNA double helix" is a prime example of a name–discovery pair. Like Abraham Lincoln, Christopher Columbus, or Cleopatra in popular history, certain scientists make such an impact that they, and their specific contributions, are modulated by collective attention and emotion, and memory for a distinct name–discovery pair becomes part of stable science memory. And like all parts of stable memory, whether discrete items or generalized categories, name–discovery pairs are caught up in a vast web of interrelations that can change over time, so the meaning to a scientific community of a particular hero can also change even though the hero has achieved scientific immortality (Kuhn, 1962/1996).

The elevation of exemplars in the stable collective memory of textbooks requires an easily identifiable source that is, however, usually oversimplified. The discoverers of the DNA double helix included British experimentalists such as biophysicist Maurice Wilkins (1916–2004) and British x-ray crystallographer Rosalind Franklin (1920–1958)—Wilkins even accepted the same Nobel Prize in Medicine or Physiology in 1962—but most laypeople only remember the two renegade theorists who synthesized data produced by others. Because the Nobel Prize committees have such a great reputation as a social hippocampus (i.e., a collective relater) for the selection and relation of "important" scientists and their discoveries, and because the Nobel Prize is not awarded posthumously, collective scientific memory has essentially forgotten Franklin, who died of complications of ovarian cancer before the prize was bestowed. Wilkins's reputation has also suffered from not being associated with the dynamic duo, "Watson and Crick." Similarly, only "Meselson and Stahl" received credit for "the most beautiful experiment on biology" (on the semiconservative nature of DNA replication), which was surely a product of the material and social resources of their respective training groups, the phage biologists, and the physical chemists at Caltech (Holmes, 2001). These name–discovery pairs stand in for years of teamwork and investigation, and students are assumed to know something about "molecular genetics" if they can cite and explain these two stand-out experiments.

Updating Winstanley's observations for American biology textbooks, physician-scholar Barak Gaster (1990a, 1990b) notes that the advanced level of a text had less influence on the incorporation of molecular genetics than the paradigm under which the author first conceived the book. Biologist William Beaver, for example, was slow on the uptake of the new, unitary genetics, because he could not reconcile it with the older, diversity model of biology encapsulated in the first edition of his textbook *Fundamentals of Biology, Animal and Plant* (1939). Even in the 1958 and 1962 revised editions of his textbook—renamed *General Biology*—he reprinted the section on genetics twice, word-for-word, in the part on plants and then again for animals. Paul Weisz, in contrast, picked up Seymour Benzer's single nucleotide DNA mutation mechanism in 1959, just 4 years after it was published, because he was already working within a molecular genetics paradigm. In general, those who began writing before the onset of the molecular revolution (around 1956) were slower to include the new material than those writing after it, because they had to make so many more changes to their existing knowledge constructs (i.e., paradigms) to integrate the new findings.

The incorporation of new knowledge may take as little as 2 years, or as long as a century, depending on existing knowledge frameworks, the revolutionary nature of the new information, and the understanding and personal decisions of textbook authors, curriculum writers, and other figures within the scientific community (Gaster, 1990a, 1990b; McIntyre, 1998). However, there has been a movement over the past 15 years or so to decrease the time it takes to publish a textbook, because textbooks are manifestly not keeping up with the journal literature (e.g., McLeod, 1991; Fung, 1997; Waldum, 2001; Smith, 2003; Azer, 2004). Even 20 years ago, physician and educator Jeremy Wyatt (1991) noted that "Textbooks, unfortunately, tend to be out of date: the volumes in bookshops today were written at least 2 years ago, possibly by authors referring back to older reviews, and up to 15 years may elapse before a genuine advance gains entry" (p. 1371). Fear of lag time in publishing has led to numerous innovations in textbook design, such as binders into which new and updated pages can be added, standard paper texts with online supplements, or completely online textbooks that can be adjusted to take into account both recent literature and student needs.

But if we take our lead from Müller and Pilzecker, who pointed out already in 1900 that the consolidation of memory takes time, then calls for ever-faster textbook production seem misplaced in light of the need to consolidate stable, time-tested collective memories for equipping the next generation of researchers with a sturdy framework against which they can judge more recent findings. As epidemiologists Fang F. Zhang and colleagues have noted about textbooks in their field:

Textbooks are an expression of the state of development of a discipline at a given moment in time. They usually do not incorporate the latest methodological and conceptual developments, but tend to present material that had been around long enough to reach some level of consensus among scholars in the field. Texts therefore rarely reflect the innovative thinking of their authors but rather the author's ability to incorporate and synthesize other people's work. (Zhang et al., 2004, p. 103)

Interestingly enough, Zhang and colleagues report that even the authors of the epidemiology textbooks they surveyed did not always include their own path-breaking work, apparently feeling the need to see whether the scientific community would accept their ideas before they published them in their own textbooks as stable collective science memories.

These examples illustrate that the stability of generalized scientific collective memory is due to its slowness to change, which derives largely from the supremely well justified concern of scientists for the establishment of the truth of the principles that they allow to become general. But slowness to change is a property that scientific collective memory shares with the memories of other types of collectives. Several authors have remarked on the "inertia" of the collective memories of groups for whom "truth" is not necessarily a primary motivator (e.g., Wertsch, 2002; Schuman et al., 2005). Seen more broadly, the slowness to change of stable memory is due to the slowness of the consolidation process itself, as new items must be related to each other and reconciled with old items and with the goals, desires, and plans of the consolidating entity. The product of the consolidation process is stable memory that is useful to its entity. James Wertsch (2002) suggests that the object of societal memory-making is the formation of a "usable past" that a whole society can draw upon as the basis for decision and action (see also Commager, 1967; Bouwsma, 1990; Woodell, 1993). In the next section, we consider the usefulness and importance of generalized collective memory to some nonscientific collectives.

Narratives and Collective Identities as Social Frameworks

Throughout this book, we define a collective simply as a group of people of any size, and we realize that larger collectives are usually composed of multiple, and possibly overlapping, subgroups or factions. Because any collective can have a collective memory, all of the various subgroups within a collective, including the collective as a whole, can have their own collective memories, and there is no requirement that the various subgroup memories be consistent with one another. In fact, subgroups within a society often define themselves according to a narrative (stable collective memory) that is not only unique to them but also contradicts the narrative of the larger society. In this section, we explore the synergistic relationship between a narrative, which defines a collective, and the collective itself, which forms the narrative.

A collective narrative serves purposes similar to the self-schema of individual memory. A collective narrative provides an explanation for a group and its place in the world and provides a group with its basis for decision and action. A narrative is also used to animate a collective identity, to draw a group together, and to keep it together. Because most groups are composed of many subgroups, this simple view is complicated by the reality that the same individual can simultaneously "belong" to several subgroups. For example, a person with dual citizenship can be an American and a Canadian, an Episcopalian, a lawyer, a Democrat, a Liberal Party member, a family man, and a member of a gardening club. Just as individual memory can be socially mediated, the memory of a subgroup can be influenced by that of the larger group of which it is a part or by the memories of other subgroups. An ethnicity is a subnational group identity that similarly requires a sense-making narrative.

According to the collected authors of the essays in *We Are a People* (Spickard & Burroughs, 2000), narrative is at the heart of (but not coterminous with) ethnic identity: This is why groups identify themselves with stories that begin, "we are the people who . . . ." They are telling the stories that describe who they are. Unsurprisingly, ethnic narratives are always multiple and therefore contested, whether within or between groups. This is because groups do not only tell stories about who *they* are, but they also develop narratives about the identities of the other groups around them. For this reason, according to Spickard and Burroughs (2000), ethnic identity is determined only in part by the ethnic group itself; other groups inevitably influence its narrative. In addition, factions within a group may assert different narratives for the entire group; this is a common source of generational, social, or other "internecine" tensions. Because narrated ethnic identities are important parts of groups' power struggles with each other for authority or status, when a group feels physically, economically, or otherwise threatened, it often turns to the discursive realm to reassert its moral position in the social order.

Narratives will be rehashed, strengthened, and occasionally modified to reduce inconsistencies. This aspect of rehearsal is fundamental to the consolidation process on individual and collective levels. A narrative may be the richest form of generalized memory, but, like any other form of generalized memory, it is shaped (and reshaped) through the consolidation of many examples and experiences. In the context of our memory model, generalized memory is formed (and reformed) as the relater repeatedly presents interrelated items to the generalizer. For a categorizer, which is the simplest form of generalized memory, the interrelated items can be presented as input–output (instance–category) pairs. For a narrative, larger sets of interrelated items may be presented as story segments or subplots. In all cases, repetition is paramount. Generalized memory cannot be formed without it.

Our model suggests that the more often a particular scenario has been played out in the experience of a group, the more likely it is to appear as a subtheme in its group narrative. A particularly appropriate example of this on a national level is the "expulsion of foreign enemies" template in Russian collective memory identified by James Wertsch (2002, 2009). As discussed previously in this chapter, Wertsch views this template as a dominant theme in Russian collective remembering. What makes this template so pertinent to our discussion of generalization is that it was formed through repetition. Although many nations have had to drive off invaders, Wertsch notes that Russia has done so with enough regularity that the basic outline of the story has become a "generalized gist." The idea that such a "generalized gist" was formed through repetition is consistent with our view that stable memory is formed through repeated presentation of information to the generalization component of the consolidation process.

Repetition can occur in the material realm, as when a group actually experiences many similar events, or in the discursive realm, as a group repeatedly recalls an event. Psychologist Henry Roediger and his coworkers (2009) have hypothesized that repeated retrieval is important for shaping collective memory. This hypothesis is also consistent with our model, but it is important to point out, as Bartlett originally demonstrated (see earlier in this chapter), that repeated retrieval leads to generalization but does not necessarily promote accuracy in recall. Just as accurate feedback (such as the correct answers on quizzes) can improve the accuracy of individual memory, Roediger et al. (2009) suggest that feedback provided by others improves collective memory as individuals in a group retrieve memories of a shared event. This suggestion presupposes that individuals within the same group can provide each other with accurate feedback. Perhaps they can. Alternatively, there may be no more potent corrective to the self-image of any group, on any level, than its interactions with other groups.

The influence of group interaction on group memory formation is apparent in the narratives that both white and black Americans have consolidated about African Americans. African Americans have long contested (white) Americans' narratives about blacks' inability to integrate into the mainstream of American society. To take but one example, in the opening of his history of blacks in America, *Black Reconstruction* (1935), American civil rights activist W.E.B. Du Bois (1868–1963) lays rhetorical waste to justifications for slavery and Jim Crow laws based on theories of racial hegemony. Rather than accept the narrative imposed on them by the larger society, Du Bois argues for a new identity for black Americans, one they can be proud to claim. In the process, he writes a very different historical narrative about American Reconstruction (1865–1877) than the one schoolchildren were still learning in the 1930s. Du Bois relates "black American" with "hard-working" and "respectable," and points out that blacks broke their backs to build the nation that refused to grant them equal citizenship. With these new attributes, Du Bois constructs a new story about blacks in America since the end of slavery and how they need to come together and assert their racial equality. He proposes an alternative narrative to the one bigots had institutionalized in American society. In a way, Du Bois is also "re-constructing" American blacks' identity and narrative.

Du Bois is perhaps most famous for his debate with fellow American civil rights activist and educator Booker T. Washington (1856–1915) over the best way to respond to racism (Franklin & Meier, 1982; Gates & West, 2000; Jonas, 2005; Bracey, 2008). Whereas Du Bois agitated for immediate change, such as opening higher education and home-ownership to blacks who had the talent and the money, Washington—"The Great Accommodator," as Du Bois called him—encouraged his fellow African Americans to get a practical education, avoid confrontation, and wait until the system had changed before demanding equal rights. This is just one example of how two different factions consolidated different identities for the larger group "African Americans" and advocated two different narratives: one of social oppression and consequent revolution and the other of education leading to slow and steady progress. These debates continue to this day. Many contemporary civil rights leaders trace their story-lines back to Du Bois, but some African Americans disagree with that narrative; they want to claim a different set of characteristics as central to their identity. The consolidation of these generalized social frameworks requires the group to situate itself in relation to other groups and to negotiate a necessarily heterogeneous collective. These examples illustrate that stable collective memories are always relational, multiple, and, indeed, changeable.

Recursion in Collective Memory Formation

As generalized collective memories, group identities are of obvious import because they bear significantly on the decisions taken by group members both individually and collectively. An existing group identity influences its own continued formation (and reformation), because it influences what the group *does*, and these actions, and their consequences, must be incorporated into the group's evolving self-concept. In addition to influencing what actually happens to the group, its identity, as generalized knowledge, more directly affects the group's collective memory formation by acting on its buffer and relater elements to influence which items of its experience are selected for long-term storage and how they are related. Thus group identity, and likewise other forms of generalized collective memory, recur to influence the ongoing processes of memory formation both of individuals in the group and of the group as a whole.

Our goal in this book is to establish that individual and collective memory formations are analogous process occurring on different levels. To compare the individual and collective levels we must separate them, but we also acknowledge substantial interaction between levels. Such acknowledgment is especially appropriate in the case of recursion, because the recursion of generalized collective memory on individual memory formation is itself a recurring theme in memory studies. Maurice Halbwachs (Halbwachs & Coser, 1925/1992) led the way by showing how social frameworks mediate remembering and memory formation. Recently researchers, mainly psychologists, have followed in his footsteps. For example, David Rubin and co-workers have shown how socially determined life-scripts organize the formation of individual autobiographical memories (Rubin et al., 1986; Rubin & Berntsen, 2003), and Baljinder Sahdra and Michael Ross (2007) have demonstrated how in-group identification strongly influences individual recall for past events. Given that stable collective memory can recur to influence individual memory formation and that individuals interacting in groups create (i.e., consolidate) collective memory, it follows that already consolidated collective memory should recur to influence further collective memory formation. One prominent argument in favor of such collective-on-collective recursion was provided by French sociologist Pierre Bourdieu (1930–2002) in his theory of collective thought and practice, *habitus*.

Bourdieu was trying to transcend objective, Marxist structuralism and outright class antagonism by developing a social theory that acknowledged economic capital not as the end-all of human striving but as the basis for social, cultural, and symbolic capital. His definition in *Outline of a Theory of Practice* (Bourdieu, 1972/1977) is classic:

The structures constitutive of a particular type of environment (e.g. the material conditions of existence characteristic of a class condition) produce *habitus*, systems of durable, transposable *dispositions*, structured structures predisposed to functioning as structuring structures, that is, as principles of the generation and structuring of practices and representations which can be objectively 'regulated' and 'regular' without in any way being the product of obedience to rules, objectively adapted to their goals without presupposing a conscious aiming at ends or an express mastery of the operations necessary to attain them and, being all this, collectively orchestrated without being the product of the orchestrating action of a conductor. (p. 72)

Habitus explains how social structures propagate through time and space: The material environment influences the immaterial desires and aspirations of the persons living in it, and these in turn direct their behavior toward (re)producing those material conditions. In rejecting the "iron cage" of the mechanistic social theories of his contemporaries, Bourdieu crafted a theory of social self-regulation that explains why nothing seems to change: Material conditions produce dispositions that reify the existing material conditions.

In writing of "structures," Bourdieu included the material, built environment as well as discursive structures such as life-scripts, group identities, social frameworks, and other social constructs that we would consider as types of stable, generalized collective memory. He clearly conceived of these structures as self-reinforcing and self-sustaining. In describing *habitus* as "structured structures predisposed to functioning as structuring structures," Bourdieu essentially describes the process of recursion. In the context of our model, "structured structures" are discursive collective memory "structures" such as social frameworks that have been "structured" by the process of collective memory formation, and "structuring structures" are these same "structures"—these beliefs, habits, practices, and understandings of the world based on past experience—influencing or "structuring" present actions including those that form (and reform) social frameworks. Bourdieu wrote of *habitus* in reference not to individual schemata but to those of whole classes: The material and discursive environment of a group recurrently influences its worldview and its practices. We see *habitus* as the collective-on-collective recursion of generalized memory onto memory formation, promoting the stability of generalized collective memory.

According to our model, recursion can operate automatically, as the generalizer directly exerts its influence on the buffer and relater, or it can operate through the entity, as the entity consults the generalizer in deciding which items should be in play in the buffer and in what ways they should be related. But the entity can operate independently of the generalizer and act according to its goals, plans, desires, and emotions. This ability of the entity underscores our view that although generalized memory exerts a powerful recursive influence, the system as a whole is not a slave to its memory—non-mnemonic entity factors can exert an equally powerful effect on memory formation. We consider these effects in the next chapter.

Conclusion

Our discussion of generalization and specialization in individual memory began with schemata, Frederic Bartlett's most lasting contribution to the scientific study of memory. Already in 1932 he recognized their subjective nature, the way they change over time, and that they are responsible for (errors in) recall. We have also explained how schemata, and other forms of generalized memory, recur to influence the processes that bring about consolidation, especially selection and relation. We have seen how the formation of generalized/specialized stable memory first requires the gathering together of many selected labile memory items and the establishment of relationships between them. Although some labile memories, modulated by attention and/or emotion, can enter stable memory as distinct items, most labile items lose their distinguishing characteristics as the generalizer extracts knowledge from the system of relationships established for them. Once created, generalized knowledge constructs can be efficiently and quickly accessed by the remembering entity, but the consolidation process that creates them is effortful and slow and can involve a great deal of contest and negotiation. Should the consolidation process be disrupted, the selected labile items that were under its consideration as potential stable memories will never be converted into a durable form (see part III for a detailed case study of disruption of the formation of a collective memory).

For much of our discussion of generalization and specialization in collective memory, we used history of science examples, because the consolidation of scientific knowledge exhibits many of the features that we broadly attribute to memory formation on individual and collective levels. Scientific data and ideas start out in a labile, contestable state as the scientific community endeavors to reproduce and verify them and relate them to each other and to already established, stable scientific knowledge. All of these processes operate under the influence of the prevailing scientific paradigm and serve to strengthen, extend, or modify it accordingly as new findings either support or undermine it. The result is a slowly changing, relatively stable body of generalized scientific knowledge that is represented in a useful, accessible, and transmissible form. The knowledge is represented largely independently of attributions to the events and individuals associated with its origination, although certain experiments, theories, and individual scientists endure in stable scientific memory as distinct name–discovery pairs. Often, these distinct memory items serve as exemplars for more general principles and are accorded pride of place in textbooks that promulgate the paradigm to the next generation of scientists.

We focused on science as a relatively straightforward example, but hopefully it is evident to the reader that the consolidation of scientific knowledge is essentially the same as the formation of class consciousness or national identity. In the larger society, the labile items are mainly political events, which are placed into a system of relationships through discourse and debate involving public figures, scholars, the media, and, to a degree made significant by their sheer numbers, members of the general public. A number of scholars have conducted similar studies with history (rather than science) textbooks, examining when and especially how certain historical details are represented (Elson, 1964; FitzGerald, 1979; Berghahn & Schissler, 1987; Frisch, 1989; Apple & Christian-Smith, 1991; Delfattore, 1992; Schudson, 1994; Tyack, 1999, 2003; Moreau, 2003). Contest and negotiation over precisely which events should be generalized and thereby stabilized for passing on to the next generation in schoolbooks is at least as old as public education (150–200 years), as various groups have competed to consolidate a version of the past consonant with their own narratives.

The generalizer/specializer is an element of particular importance in our consolidation model for two main reasons. First, the generalized knowledge constructs created in this last stage of consolidation are, in some sense, the end product of the entire process. The buffer, as the first stage, is necessary for temporarily holding labile memory items. The selector/relater, as the second stage, is central because it places selected items into a system of relationships. The generalizer, as the third and final stage, extracts the knowledge inherent in this system of relationships and represents it in a ready, usable, and, for collective memory, transmissible form. The great achievements of individual and collective memory consolidation, including classifiers, categorizers, schemata, paradigms, social frameworks, national identities, and cultural narratives, are all forms of generalized stable memory.

The second reason the generalizer/specializer is of particular importance, in seeming contradiction to the first reason, is that it is *not* the end of the consolidation process, because it can recur to influence the consolidation of new, generalized memory. Thus, memory consolidation is a dynamic, ongoing process, which continually re-creates generalized memory constructs, which are stable but not rigid, enduring but not immutable.

The fourth element of our consolidation model is the remembering entity, which contains the other three elements. Like the generalizer/specializer, the entity can also influence the consolidation process. We describe the entity and its influence on consolidation in the next chapter.

8 Influence of the Consolidating Entity

The fourth component of our three-in-one model is the consolidating entity. In the context of individual memory consolidation, the entity is a person; in that of collective memory consolidation, the entity is the particular collective in question (e.g., family, group, community, society, nation). The entity arises from the synergistic interaction between its many components, and it is all encompassing. It not only contains the other three elements of our consolidation model (buffer, relater, and generalizer) but also, by definition, contains every other element of its being, whether mnemonic or non-mnemonic. Thus, in addition to declarative memory, which is the form of memory we consider in this book (see chapter 2), the entity's contents include procedural memory, which is the form of memory we do not consider. The non-mnemonic contents of the entity include the cognitive elements responsible for thinking and planning, for mediating a sense of coherence, and for the emotions. These processes are themselves extremely complex and are hard enough to describe on the individual level, much less on collective levels. The entity is all the more complex because it arises out of the interaction of its already complex components. We will not attempt to describe the all-encompassing entity in its entirety. Because our focus is strictly trained on memory consolidation, consideration of the remembering entity as a whole is beyond our scope. In this chapter, we concentrate on the influence of the entity on the consolidation process, which occurs mainly, but by no means exclusively, through the relater. This chapter begins with a genealogy of the concept of the entity's effects on consolidation. We then offer examples of individual and collective entities to demonstrate how non-mnemonic factors also influence the memory consolidation process.

The Role of the Entity in Our Model

A productive way to explain the role the entity plays in consolidation is to review our three-in-one model and the history of the scientific understandings of memory formation that led up to it. The basic trope for this discussion is the "box." We use boxes as placeholders into which we can put abstract concepts, processes, or objects, which can stand either for real memory elements or for metaphors that help us explain the nature of real memory elements. For example, the first box in our three-in-one model is the buffer. We can place in it the abstract concept of a memory buffer; or a process that temporarily stores memory items; or a part of the brain (e.g., frontal cortex) or a cultural tool (e.g., newspaper) that physically holds memory items temporarily; or a metaphor for a buffer, such as a leaky bucket. All of the different things we place in the first box correspond to buffered storage, but each allows us to elucidate a different aspect of what is a complex and multifaceted element of the consolidation process.

With this understanding of boxes, let us begin with a one-box model of memory. Hermann Ebbinghaus's (1885) list-learning paradigm drew on a model of memory that essentially consisted of one box to store memories. The nonsense words read from a list either entered the subject's memory or they did not. Remembered words could be forgotten, but memory was stronger for the same words read repeatedly. Despite its simplicity, this single-box model comes with powerful capabilities: It explains basic learning, retention, and forgetting, and "savings" (*Ersparnis*), or the memory that is "saved" between the first and second exposures to a list. What the single-box model could not explain, however, is interference: Information learned earlier can inhibit the retention of information learned later, and vice versa (these phenomena are termed proactive interference and retroactive interference, respectively; see Hebb, 1972).

Considering the findings on learning interference, Georg Müller and Alfons Pilzecker (1900) concluded that there must be some "perseveration" period during which memory items are held in a disruptable state while they are being "consolidated." They therefore added what amounts to a short-term holding box (i.e., buffer) in front of the long-term storage box. According to their conceptualization, memories are retained in the holding box for the short term before they are transferred to the storage box for the long term. The idea of a physical "transfer" from one box to another is a metaphor for a real, neurophysiologic transformation of memory from a labile to a stable state. The metaphor uses visual imagery to convey the concept that memory items are disruptable (labile) until they become permanent (stable).

The two-box model nicely explained the basic temporal properties of retrograde amnesia, in which memory is lost for recent events but not for events that occurred further back in time (W. McDougall, 1901; Burnham, 1903). The two-box model of Müller and Pilzecker metaphorically accounts for retrograde amnesia findings as the disruption of the movement of memories from the first to the second box. Like the one-box model, the two-box model can have various instantiations. For instance, both consolidation and Ebbinghaus's forgetting curve can be simulated using the "two-bucket" model, in which memory, represented metaphorically as a fluid, flows from the first, more-leaky bucket to the second, less-leaky bucket (see chapter 2). The two-box model has also supported a number of neuroscientific explanations of memory formation. For instance, Donald Hebb (1949) in his dual-trace theory explained memory consolidation in terms of the wiring of neural circuits. Now the first box corresponds to reverberation among weakly connected neurons arranged in circuits, and the second box corresponds to the establishment of strong synaptic connections between the neurons. To represent Hebb's dual-trace theory, both boxes could stand for the same neural structure, but memories are stored in terms of neural activity in the first box and as synaptic weight changes in the second box. In that it describes possible neurophysiologic mechanisms, the dual-trace theory is hypothetical (still incompletely proven) but literal rather than metaphorical.

One of the most successful applications of a two-box model is McClelland and colleagues' (1995) computational neuroscience model composed of a Hopfield network (first box) and a multilayer perceptron trained by back-propagation (second box) (see figure 2.7). This model is both literal and metaphorical: literal, because the boxes hold artificial neural networks that can be implemented on computers and actually perform their intended functions; and metaphorical, because the neural networks are also meant as models of real neurophysiologic processes. When implemented on computers, the neural networks operate on patterns that represent memory items. A pattern is simply a particular arrangement of activity over the model neurons (units) that compose a neural network. For example, the pattern 0 0 1 0 1 could be represented by a Hopfield network composed of five units, each of which could be inactive (at level 0) or active (at level 1). A five-unit Hopfield network could represent 32 different binary patterns (simulated memory items) but, due to inherent limitations, it could reliably store and recall only a small fraction of them.

A Hopfield network has one layer of units, but that layer can be divided up into two or more segments. A multilayer perceptron has explicit input and output layers and one or more intervening (i.e., hidden) layers (see figure 2.7). A Hopfield network has a limited capacity, so it can only hold a relatively small number of patterns at any given time. In contrast, a multilayer perceptron has a practically unlimited capacity, because it can learn to transform a potentially very large number of input patterns into a smaller number of output patterns. A Hopfield network can repeatedly present verbatim patterns, but it cannot generalize. In contrast, a multilayer perceptron can learn to generalize, but it needs verbatim patterns repeatedly presented to it. The model of McClelland and colleagues combines both network types by using a Hopfield network as a buffer and a multilayer perceptron as a generalizer. Specifically, the Hopfield network repeatedly presents patterns, as input–output pairs, for the multilayer perceptron to learn using back-propagation to correct its recall errors (see chapter 2; for background see Anastasio, 2010).

Each box in the two-box model of McClelland and colleagues stands simultaneously for a concept (labile-verbatim memory or stable-generalized memory) and a process (buffering or generalization). Not only does this model produce learning and forgetting curves that are consistent with observations on interference, but also it explicitly describes the rehearsal process by which memory items could become stable records from labile traces. What this model does not address is how the input–output pairs are associated. Specifically, the model can easily *represent* the input and output patterns as the first and second segments of the verbatim patterns held by the Hopfield network. What the model cannot do is *determine* which output pattern should correspond to each input pattern. In other words, although it explains memory formation quantitatively and temporally, the model of McClelland and colleagues does not explain it qualitatively in terms of either memories' contents or the larger contexts of the consolidation process.

This is where our third box comes in. We added a box between the buffer and generalizer boxes that we call the relater. Like all of the elements in our three-in-one model, the function of the relater is intended literally to correspond to a component of the actual process of memory consolidation. The relater relates memory patterns one to another, forming webs of relationships that can be used to organize patterns (items) from the buffer into input–output pairs for presentation to the generalizer. The relater corresponds to the hippocampus in the brain of an individual, and it could, in principle, establish relationships automatically.

Howard Eichenbaum (2006) has suggested a temporal basis on which the hippocampus could build these relationships. To take some examples from everyday life, a person could experience the sequence: table–salt–shake–flavorful food, and then could experience the sequence sidewalk–salt–spread–melted ice. The individual items would be held temporarily in the buffer. The hippocampus (relater) would keep track of their temporal relationships and use those to form webs of associations, from which it could extract the input–output pairs it presents to the generalizer. The relater is not limited to the original temporal sequences on which its web of connections is based; it can join them via common elements and then traverse ("surf") them going forward, in reverse, one item at a time, or in leapfrog mode. To continue our example, the relater could join the two sequences at "salt." Then, going in reverse, it could form the two pairs shake–salt and spread–salt, from which the generalizer could learn that shaking and spreading are both actions that can be performed with salt. Similarly, in leapfrog mode, the relater could form the pairs salt–flavor and salt–melt, from which the generalizer could learn that salt is associated both with gastronomy and avoidance of liability. While it is easy to see how this process (buffer–relater–generalizer) could form knowledge constructs such as associators and classifiers, it is reasonable to assume that scaled-up versions could also form generalized memories such as schemas, narratives, and identities.

The three-box model, composed of a buffer, a relater, and a generalizer, could account for memory consolidation by automatically holding, interrelating, and generalizing a set of labile items into a stable knowledge construct. Because the relater corresponds to the hippocampus, and because it is the central element in the three-box model, the model also nicely accounts both for the retrograde amnesia and anterograde amnesia (inability to form new stable memories) that occur after damage to the hippocampus. What the three-box model cannot do is account for memory inaccuracies. Of course, we could assume that any one or all of the elements sometimes fail to function properly, and this would result in memory errors, but random malfunctioning would not explain the many systematic inaccuracies that have been observed. Some of these inaccuracies, as Bartlett (1932) described, can be attributed to the particular characteristics of an individual, such as his profession or ethnicity.

Cognitive psychologist Ulric Neisser provides a nice example of individual entity effects. Neisser (1981), in an intriguing case study, examined the statements of John Dean concerning the Watergate scandal, in which Dean was a participant. Neisser found that Dean got the facts mostly right, but he inaccurately overstated the importance of the role he played over the course of those infamous events. Assuming that his hippocampus acted as an automatic relationality element would not explain why John Dean would "remember" that he played a larger role in Watergate than he actually did. Discussion of why a person might "want" to construct a personal narrative that overplays his role in a scandal is beyond our focus. At the same time, we must take into account that non-mnemonic factors such as desires, goals, plans, emotions, and a sense of coherence concerning the self and its place in the world can all influence the memory consolidation process. This is where the entity comes in.

The entity is the fourth box in our model of memory consolidation, but it is a special box. It represents the entire remembering being, so it encompasses all of the components of that being including the other three boxes that mediate memory consolidation proper. We place the three boxes representing the core consolidation process (buffer–relater–generalizer) within the entity box, creating a three-in-one model (see figure 4.2). The entity can access the contents and affect the function of the other three boxes. When the entity remembers, it can use any or all of the other elements, taking labile items from the buffer and stable items from the generalizer, and it can use the relater to relate these items flexibly. Likewise during consolidation, the entity can choose which items to attend in the buffer and which items to draw from the generalizer, and it can direct the way the relater relates those selected items. Entity effects can explain, on the individual level, why John Dean and Richard Nixon would consolidate different versions of the Watergate story from the same set of facts. On the collective level, entity effects can also explain why Greek and Turkish Cypriots would consolidate different versions of the history of Cyprus (see chapter 6) from the same series of events. It is not that their buffers or relaters or generalizers work differently that accounts for the differences in memory, but that each entity influences the consolidation process according to its own, unique, often selfish purposes.

The influences that the entity exerts on the consolidation process can be mnemonic or non-mnemonic. The mnemonic influences come mainly from the generalizer, which is as much a part of the entity as any other part of its being. But the generalizer can exert an automatic influence on consolidation that could operate in a reduced model composed only of the buffer, relater, and generalizer (independently of the rest of the entity). At the most basic level, the generalizer automatically downloads to the buffer items that have already been consolidated, so that the relater can relate them with new items. This mechanism is essential to the consolidation process, because it ensures that already consolidated items will not be overwritten by new items but instead will be intermingled with them and provide the context in which the new items are consolidated. In this way, already consolidated, generalized memory can automatically recur to guide the consolidation of new memory.

Although automatic recursion of generalized memory probably occurs in the intact entity, the omnipotent entity can also direct this recursion and determine, for example, which items specifically are downloaded from the generalizer, which are selected from the buffer, and how they are related. The entity can use generalized memory constructs, including its self-schema or identity narrative, as its basis for deciding how to direct the consolidation process. Thus, the influence of generalized memory on consolidation is recursive, whether it occurs automatically or through the entity. What the entity adds are non-mnemonic factors such as goals, desires, emotions, and a sense of coherence. Our entity is similar to "the working self," which psychologists Martin Conway and Kit Pleydell-Pearce (2000) see as the agent that directs the formation of autobiographical memory. The working self maintains coherence among current goals and draws from already established autobiographical knowledge to determine which new events are consolidated into autobiographical memory (Williams & Conway, 2009). More generally, the entity, whether an individual person or a collective of any size, can influence the consolidation of its own declarative memory. Once again, a detailed description of the all-encompassing entity is beyond the focus of our study, but we do need to account for its effects on consolidation. We spend the rest of this chapter doing that on individual and collective levels.

Individual Entity Effects

With the review in the previous section, we are ready to provide some concrete examples of entity effects on memory consolidation. To reiterate, any factor that affects consolidation that is not strictly memory falls under the heading "entity effects." One such effect is coherence, by which an entity strives to "make sense" of the events it remembers. Bartlett (1932) recognized this property early on and referred to it as an "effort after meaning." More recently, psychologists Michael Ross and Anne Wilson (2003; see also Wilson & Ross, 2001) showed that individuals, who mostly see themselves as improving over time, tend to revise their autobiographical memories so that their past self, as they remember it, is worse than their current self. This "derogation" of the personal past is a memory distortion, but it makes autobiographical memory more consistent with the theme of personal betterment over time. Especially with regard to the maintenance of a favorable and continually improving self-appraisal, these results suggest that individual entities are more tolerant of memory inaccuracy than of incoherence.

Other entity properties that can influence individual memory consolidation include goals, plans, desires, and emotions. As discussed in chapter 2, emotion is known to modulate or "tag" certain memories for consolidation as distinct items. The study of the effects of emotion on memory blurs the line between individual and collective memory, mainly for two reasons. First, in order to find common features among the emotionally modulated memories of individuals, researchers often focus on the memory of many individuals for the same shared (i.e., social) event. Second, social frameworks can influence individual emotion as well as individual memory (e.g., Halbwachs & Coser, 1925/1992). We try to avoid individual–collective interactions so that we can keep the levels (artificially) separate as we draw an analogy between them. Here, in the case of entity effects, we acknowledge a strong individual–collective interaction, but we still clearly see the influence of emotion and other entity effects on memory consolidation in individuals.

Although "arousal" cannot be measured directly, the physiologic state known as "arousal" in individuals can be detected indirectly as changes in heart rate, breathing rate, blood flow patterns, the levels of circulating hormones, and other measurable quantities. As James McGaugh (2000) has explained, arousal has very definite effects on memory consolidation. Emotionally arousing events—be they extremely happy, surprising, enraging, disappointing, humorous, or frightening—cause the adrenal glands to secrete the hormones epinephrine (also know as adrenaline) and (in humans) cortisol. These hormones activate, among other things, the amygdala, a small brain structure that lies in front of the hippocampus. These neuroendocrinologic events, which contribute to the experience of "emotion," strengthen the traces of memories that are associated with the emotionally arousing events and can speed up the time course of the consolidation process.

The effects of emotions on memory consolidation are clearly observed in the context of so-called "flashbulb" memories. For Americans, common (and shared) subjects for flashbulb memories include the assassination of President John F. Kennedy (November 22, 1963), the attempted assassination of President Ronald Reagan (March 30, 1981), the *Challenger* disaster

(January 28, 1986), and the terrorist attacks on September 11, 2001. No mere academic theory, the idea of flashbulb memories has also permeated lay conceptions of memory. For instance, in its November/December 2003 issue, *Reminisce: The Magazine That Brings Back Memories* published four pages of reader-submitted anecdotes about "The Day We Heard about JFK." One of them begins, "Most people remember where they were on that Friday afternoon when they heard President Kennedy had been shot. I remember where I was on that Friday night. I was in the Army, with the 97th Signal Battalion in Germany, where it was 7 hours later" (Schroeder, 2003, p. 30). He goes on to describe what he was doing (cleaning his gear), how he heard the news (Armed Forces Radio), and who was with him (the guys down the hall were playing cards). This great attention to detail has been described as characteristic of flashbulb memories.

Psychologists Roger Brown and James Kulik first reported the flashbulb memory phenomenon in 1977, suggesting that the brain has a special mechanism for consolidating emotionally charged events in a durable fashion as distinct memories with incredible detail. Since then, psychologists have heavily debated and modified the theory. A major criticism is that, rather than being indelibly inscribed in brain tissue, as originally advertised, flashbulb memories appear to decay just like "regular" episodic memories (Curci et al., 2001). Another criticism is that flashbulb memories are notoriously inaccurate (McCloskey et al., 1988; Neisser & Harsch, 1992; Loftus, 1997; de Rivera, 1998; Hyman & Loftus, 1998; Curci & Luminet, 2006). Social psychologist Catrin Finkenauer and colleagues (1998) therefore have suggested a particularly useful modification, the "emotional–integrative model," which attempts to find a middle ground for flashbulb memories somewhere between photographic and fading.

According to the emotional-integrative model, surprising events first cause abnormal emotional states that *directly* affect consolidation via modulation, as the neuroimmunologic research reported above would suggest. The highly emotional nature of these events causes individuals to discuss them, leading to a second, *indirect* effect via rehearsal, which is the more general process in consolidation. This take on memory consolidation blends internal factors (individual hormonal physiology and emotions) with external ones (social situations), so it involves the sort of individual–collective interaction that we acknowledge but generally avoid in this book in order to stress the analogy between levels. But it emphasizes the effects of individual emotion on individual memory consolidation, so it is precisely what we would expect from the "entity" concept in our three-in-one model.

Flashbulb memories illustrate another entity effect on individual memory, which Bartlett presaged in his repeated recall studies. Bartlett (1932) observed that a subject's attitudes and interests, as well as his occupation, ethnicity, and nationality, could significantly influence his memory and recall. In accordance, Brown and Kulik (1977), the originators of the flashbulb memory idea, reported that African Americans remembered the assassinations of civil rights leaders better than did European Americans. More recently, it was reported that Americans remember the events of September 11, 2001, better than do Europeans (Curci & Luminet, 2006). These effects could be due to the influence of social frameworks, in which case they would be caused by the generalizer element rather than by the entity, but other results suggest that non-mnemonic entity effects are also at work. Psychologists Baljinder Sahdra and Michael Ross (2007) have shown that among ethnic Sikhs and Hindus, those with strong in-group identity remember fewer incidents of violence toward the out-group (respectively Hindu or Sikh) than those with weak in-group identity. Whereas ethnicity and nationality are social frameworks, the degree of in-group identification is an individual entity property. These results suggest not only that social frameworks (a collective generalizer property) but also in-group identification (an individual entity property) can influence which items are selected for consolidation.

Collective Entity Effects

As we have already mentioned, an in-depth analysis of entities qua entities is outside our scope, and so are analyses of the properties of entities, except, of course, for the property of memory consolidation, which is our main focus. For other entity properties, we can adopt the same strategy we used in our discussion of collective memory qua memory (see chapter 3), which is that individual entities are assumed to possess certain properties, such as goals, plans, desires, and emotions, and that their collective counterparts also exist as more than the sum of (or at least different from) the combined properties of the individuals composing the collective. Mob rule provides a ready example of collective desire, as it is different from the desire that any individual would have if separated from the group and from the combination of those individual desires. Similarly, collectives can plan to achieve goals that would simply seem impossible to isolated individuals. The idea of collective emotion is more elusive but can be thought of as an impulse that drives a group to collective action, just as an emotion motivates an individual into action.

Collective emotion can motivate interaction among the members of a group. Using an innovative combination of personal audio recorders and computer analysis of online Web logs (blogs), psychologist James Pennebaker and his co-workers have studied the responses of individuals in society to tragic events, including the terrorist attacks of September 11, 2001, and the death of Princess Diana (Stone & Pennebaker, 2002; Mehl & Pennebaker, 2003; Cohn et al., 2004; Pennebaker & Gonzales, 2009). Pennebaker's team has found that, relatively soon after a tragic event, people engage mainly in dyadic interactions. The authors interpret this to mean that after the initial shock, individuals need to develop their own personal stories of a tragedy with a few "trusted others." In the weeks and months after a tragedy, people gradually re-emerge into the larger collective. Word-usage analysis shows that they tend to use "we" more than "I," indicating a more group-oriented attitude. Also, negative word usage initially rises and then falls back to baseline, whereas positive word usage initially falls but then rises above baseline.

This is all taken to mean that a tragedy initially makes people negative and withdrawn, but the initial disengagement gradually switches to a positive broadening of outlook, increased interaction, and a greater sense of community. Pennebaker and Gonzales (2009) consider this biphasic pattern of reemergence following withdrawal as an early sign that the tragic event that precipitated the response will ultimately enter collective memory as a distinct item. An even earlier predictor is the amount of emotion the event elicits, which is a sign of its significance to the society. Thus, Pennebaker and his research group paint a picture of the influence of emotion on the consolidation of collective memory: Emotion elicited by a tragic event causes a biphasic response characterized by initial withdrawal and subsequent re-emergence into society, and the resulting community-wide processing of the event leads to its eventual consolidation into collective memory.

This intriguing idea, which Pennebaker and coworkers describe on collective levels, is similar to the emotion-triggered response Finkenauer and colleagues (1998) describe on the individual level. According to their "emotional–integrative model" (discussed earlier), emotion immediately tags an arousing event for individual consolidation, and its highly arousing nature later leads the individual to discuss it with others, whether the community shared the event or not, and this discussion enhances consolidation. Together, Pennebaker and Finkenauer's ideas illustrate possible mechanisms by which emotion could modulate collective and individual memory. They are consistent with our view that the entity can influence the consolidation process in many ways, and specifically through emotional modulation.

Another way that the collective entity can influence collective memory formation is through its identity claims. James Wertsch (2009) has suggested that collective memory formation is often undertaken to provide a "usable past that serves some identity project" (p. 123). It is important to point out that the pursuit of an "identity claim" or the servicing of an "identity project" are goals, which are entity effects and so are different from an "identity" itself, which is a form of stable, generalized memory. In chapter 7, we noted two reasons why a generalized memory, such as an identity or self-schema, is slow to change. One reason is that the consolidation process that forms an identity takes time. The other reason is that an identity, once formed, can recur to influence further consolidation by downloading old (already consolidated) items to the buffer where they mix with new (unconsolidated) items and are interrelated by the relater. These are nonentity (pure memory formation) processes that tend to maintain the stability of an identity (or any other form of generalized memory) by preventing the overwriting of old by new items and by ensuring that old items are interrelated with new items.

Now we can add a third reason why a generalized memory, if it is an identity, and especially a collective identity, is slow to change: An identity is tied to the identity claims of the entity. In chapter 7, we noted that stable memory is not unchangeable and that collective stable memory can be changed from within the group or, more often, as a consequence of interactions between groups. But its identity narrative is what defines a group as a group. A collective entity's identity (a society's "usable past" in the form of an identity narrative) is slow to change because the entity will resist changes to its identity that violate its identity claims.

An insightful example of how a group's identity project and claims can influence the formation and stability of a group's identity is provided in Peter Novick's *The Holocaust in American Life* (1999). Novick argues that the reason the Holocaust became so important in American collective memory when it did, in the 1970s, is because American Jews were searching for an identity during that period. Writing of collective memory in general, Novick hypothesizes that collectives "center" certain memories that are considered to be defining or central to an established or developing collective identity. He suggests that the Holocaust has filled the need of American Jews for a "consensual symbol." Yet Novick finds the Holocaust to be a misplaced center of identity because it is so far removed from most Americans' experience and because it is uncontroversial (in America and in most of the rest of the world), so it is uncontested and therefore not part of the ongoing process of (re-)creating American identity. We can recast Novick's concern in the parlance of our model: The fact that the Holocaust is uncontested in America means that it is not, as an item, being interrelated with new and other old items in the continual process of reshaping stable collective American memory.

The example of the Holocaust in America highlights the fact that collective entities exert influences on collective memory formation based on their goals, such as identity projects. History abounds with examples of collective entities striving to shape and reshape the memory of their collective, or those of other collectives, to suit their own purposes. We will pick up this thread again in part III. In the next two sections, we return to the specific example of science to explore some of the non-mnemonic effects that operate on scientific memory formation.

The Laboratory as Entity

Laboratories—microcosms of "science"—are complex and mnemonic entities. Since the 1960s, sociologists and historians of science have studied "the laboratory" as a unique social institution. For instance, Bruno Latour and Steve Woolgar (1979) sought to make the laboratory "strange" when Latour did anthropologic fieldwork in Jonas Salk's laboratory in La Jolla, CA. Only by attempting to approach with fresh eyes what had become a familiar conglomeration of people, instruments, test tubes, animals, and specimens did they believe that they would be able to discern what *really* happens in the "science" that takes place in laboratories every day. One of their novel findings was that the most important place in a laboratory is not the bench but the desk, where the primary investigator not only coordinates the movements of the members of the laboratory and synthesizes their results, but also secures funding, material, and personnel. The primary investigator serves as a conduit between the "inside" and the "outside" of the lab, communicating and negotiating with other laboratory directors and, especially, with funding agencies. The importance to the practice of science of writing articles, progress reports, books, and (especially) grant proposals points up the fact that laboratories are not just places where scientists *do* science; in addition to the practice of science there is also a discursive side, which is as important to the practical needs of the laboratory as to the consolidation of scientific knowledge.

We have already discussed the most obvious manifestation of collective science memory, paradigms. As we detailed in chapters 4 and 7, a paradigm is the generalized knowledge construct that is shaped by the collective process of memory consolidation that occurs within a scientific community and that in turn shapes what the scientists think, do, and see, what questions they ask, how they pursue those questions, and how they interpret the answers. If journal articles constitute the usual form of labile collective science memory, then paradigms tend to be consolidated in textbooks, the most common cultural tool for the communication of stable collective science memory. While the establishment of the "truth" of the items presented in textbooks is the overriding desire of the scientific entity in influencing the formation of this particular collective memory structure, other less lofty entity effects also exert themselves.

Perhaps the most benign entity effect on scientific textbook production is marketing. Publishers exist to sell books and are justifiably reluctant to agree to produce a book that is unlikely to sell. Yet small-run, very specialized scientific textbooks are produced all the time and are purchased at high prices by specialists in the field, who justify the cost as a research expense. On the flip side, the most malicious entity effect on scientific textbook writing is censorship. A well-known American example concerns the attempt to censor the theory of evolution from high-school biology textbooks (e.g., Skoog, 1984; Rosenthal, 1985). Obviously, this type of censorship has little effect on textbooks produced specifically for scientists. As explained in chapter 7, much of the time lag between the discovery of a new fact or elucidation of a new principle and its appearance in textbooks is due to the "contest and negotiation" associated with the establishment of its validity and its integration with existing knowledge. The scientific method itself ensures that any subjectivity that creeps into the process is eventually overcome. Shakespeare could have written "at the length truth will out" in reference to science textbooks. Scientists, for their part, understand the time required for scientific facts to become established, but this is not necessarily also the case for other groups within society.

Scientific versus Political Collective Memory Consolidation

We can gain further insight into the effects of the entity on collective memory formation by comparing the consolidation process between scientists and makers of public policy in, specifically, the United States. Scientists, as a collective entity, are justifiably preoccupied with truth and are relatively unconcerned with the time required to establish it (see chapter 7). In contrast, policymakers are often more concerned with the quickness of consolidation than with truth per se, as they have decisions to make, budgets to balance, and constituents to appease. Even while American scientists are hammering out the details of some potentially stable memory, American politicians may seize upon a particular fact or theory and use it as the justification for a piece of public policy.

In a précis for a "Third Wave of Science Studies," sociologists of science Harry Collins and Robert Evans (2002) compare the amount of time it takes for some new piece of information to be accepted (i.e., consolidated) among the "core group" of scientists working on a problem, on the one hand, with the time it takes for that information to be applied "outside" the realm of science, in politics, on the other hand. Defining policymakers as that part of society concerned with creating the public policy that affects all citizens (scientists and nonscientists alike), Collins and Evans find that, when it comes to reaching a working consensus about some new scientific knowledge, the policymakers are much faster than the scientists. The authors attribute the discrepancy to the fact that "[t]he consumers, as opposed to the producers, of scientific knowledge have no use for small uncertainties" (Collins & Evans, 2002, p. 246). Policymakers often need to make qualitative decisions "before the dust has settled," as Collins and Evans put it, because political action would be stymied if policymakers had to wait the years it can take until scientists become confident consumers of their own knowledge.

Although we assume their consolidation processes are structurally the same, scientists and public officials differ in their goals as consolidating entities. Scientists want to be as sure as they can be that a new fact or theory is true before they allow it to become part of general scientific knowledge. Sometimes, however, scientists do not get a chance to reach their own consensus before policymakers come to some agreement about the new knowledge among themselves so they can consolidate it into a law or regulation. Scientists and policymakers may even consolidate more or less the same stable memory, but they do so at vastly different rates. We can conclude that the needs, desires, and goals of the collective as an entity can influence not only the content of stable collective memory but the time course of consolidation as well.

The example of scientists and policymakers also illustrates, quite clearly, that subgroups can influence each others' processes of collective memory consolidation. Sociologists of science such as Bruno Latour and Steven Woolgar (1979; see also Latour, 1987) have demonstrated since the 1970s that it is impossible to protect "science in public" from political influence: This is because politics has already entered the laboratory in the choice of

research topic, funding, and conclusions drawn. Science is rarely independent from politics (especially in a country whose post–Second World War
federal dollars birthed Big Science), but politics is not necessarily antithetical to science; many different political views encourage many different
kinds of science. Political exigencies, in other words, are yet another non-
mnemonic entity factor that can influence what labile items are attended
in the collective memory buffer, what is selected and how it is related, and
under what generalized categories to consolidate new stable collective
memories. Politics, broadly construed, can influence the consolidation of
scientific knowledge from outside the scientific community, because policymakers can influence the course of science, mainly through funding
priorities. Politics can also exert an influence from inside the scientific
community, as scientists, like most other people, simultaneously can be
members of multiple collectives, each with their own collective goals and
desires.

Conclusion

In conclusion, the process of memory consolidation, although self-
referencing via recursion, cannot be contained wholly inside itself. Factors
outside the process—entity effects—influence what is referenced and how.
As Bartlett (1932) demonstrated long ago, certain attributes of an individual, such as his personality, profession, ethnicity, or nationality, can
affect memory consolidation, and these early observations have received
more recent support. For example, Canadians altered stories in serial recall
to favor their home country but not another country (specifically, Australia) (Blatz & Ross, 2009). Many other examples on both individual and
collective levels were cited in this chapter. Findings such as these demonstrate the effects of the entity on the consolidation process in individual
and collective memory.

We considered another, collective example in chapter 6, in which we
described how Greek and Turkish Cypriots each consolidated the same set
of facts into different narratives, as displayed in their respective museums.
Our analysis in this chapter suggests that the Greeks and Turks consolidated differently because, as collective entities, they had different goals
and different emotions, and these divergent, non-mnemonic factors drove
their respective narratives in different directions. Although our focus in
chapter 6 was on the relationality element (relater), the example of the
Cypriots is also one of the influence of the consolidating entity on relationality, and so on the overall consolidation process. One could no doubt

find other examples in which desires, goals, plans, and emotions all operate as entity effects and exert an influence on individual or collective consolidation.

Our analysis in this chapter is intended mainly to acknowledge entity effects on consolidation, not to describe the remembering entity in detail. It is also intended as a lead-in to part III, where we examine the effects of a determined entity that strove to influence the memory consolidation of a large collective. It did this not so much by selecting specific, new items from the buffer, or by deciding which old items to download from the generalizer, although those influences occurred as well. In the case study we treat in detail in part III, the entity in question influenced the process of collective consolidation through an essentially complete destruction of the relationality element. By removing the collective relater, this entity performed a social hippocampectomy. The aftereffects, which we document in part III, support a prediction derived from our model of memory consolidation.

III Disruption of Consolidation

9 Collective Retrograde Amnesia

Here in the last third of the book, we use our model of memory consolidation to derive a hypothesis and present evidence in support of it. The case study presented here in part III hypothesizes that collective retrograde amnesia results from a social trauma that specifically disrupts the consolidation of collective memory in a society. Retrograde amnesia would be any loss of memory for events that occurred before the insult that caused the amnesia, and it may be induced in individuals through various neurologic or psychologic pathologies (see chapter 2). Retrograde amnesia that results specifically from interference with the process of consolidation has a characteristic profile: Memory is lost for relatively recent events that were in the process of consolidating, but more remote, already consolidated memory is preserved (see figure 2.6). Our hypothesis is that after a social trauma that disrupts the process of collective memory consolidation, a collective retrograde amnesia that fits this profile should be observed.

To test this hypothesis, we will compare the collective memories of two peoples and their immediate descendants. Both groups are of the same Chinese cultural origin, but one has experienced a specific social trauma that the other has not experienced. The trauma affected the "social hippocampus," that part of the society that establishes the relationships among labile items that are eventually consolidated into a usable, accessible, and stable form. The trauma in question is the Chinese Cultural Revolution, which affected mainland Chinese but spared Chinese populations in such places as Táiwān and Northern Thailand.

To view Chinese collective memory over a long time span, our analysis focuses on religion, where the events preserved by tradition occurred centuries ago, and literature, with a focus on the novel, which is a more recent cultural form. Our main finding is that the mainland Chinese experienced collective amnesia for literature and literary figures that were well represented as labile memory items at the time of the Communist takeover but did not experience collective amnesia for religious personages, tenets, and practices, which had been consolidated generations before and were already well represented as stable memory items. We gauge this memory loss by comparison with non-mainland populations, whose experience of the development of Chinese culture was relatively unaffected by the Cultural Revolution. The "relaters" in these non-mainland societies, who composed their respective society's "social hippocampi," were not imprisoned or killed or otherwise prevented from performing their memory-forming roles. The identification of the hypothesized retrograde-amnesia pattern after a social trauma directed specifically at the relaters in a society provides strong support for the actual existence of the process of collective memory consolidation.

Our methodological challenge in part III is to compare the declarative collective memories of two geographically and politically separate but culturally related peoples. We will do this mainly by means of comparing their memory of religion and literature, as conveyed in writing (books, including novels), and by what they know, as described in the reports of anthropologists and other observers, about collective memory structures such as religious rituals, traditional performances, architectural works, and other cultural tools. In individual psychology, there is a distinction between declarative (knowing that) and procedural (knowing how) memory (Squire & Zola, 1996; see also chapter 2). In this regard, it is important to point out that we consider only *declarative* collective memory, which is what individuals, in interaction with each other and with cultural tools, can *declare* they remember either verbally or in writing. We do not deny that collective procedural memory could exist, but we do not consider such memory in our comparison. Declarative and procedural memory are also referred to as explicit and implicit memory, respectively (Schacter, 1987). We emphasize that our comparison between mainland and non-mainland Chinese involves explicit expressions of collective declarative memory.

Our comparison between mainland Chinese, on the one hand, with Chinese in Táiwān and Northern Thailand, on the other, will not be exhaustive. By focusing on two aspects of one culture, Chinese religion and literature, we hope to provide evidence enough to convince scholars that collective retrograde amnesia can be caused by specific forms of social trauma and to inspire them to look for it within their own areas of expertise. It is important to point out that we are not interested in memory of the social trauma itself or in the effects on collective memory of social trauma in general. Instead, we are interested in the effects on collective memory consolidation (formation) of social trauma that specifically targets the relaters in a society. (The memory of social trauma itself has been an important theme in humanistic studies, and we discuss it briefly in relation to our model in chapter 12.)

Our analysis takes into account a series of Communist movements in mainland China from 1950 to the mid-1970s, in particular the Cultural Revolution from 1966 to 1976. The two peoples we compare are the mainland Chinese, who experienced these Communist social programs, and the Chinese of Northern Thailand and of Táiwān, who did not. The Chinese in Northern Thailand reside in roughly 67 villages. They are those (and now mainly the descendants of those) who were associated with the Guómíndǎng (Chinese Nationalist Party, or KMT in the Chinese spelling system adopted in Táiwān) and who emigrated to Northern Thailand from China around roughly 1950, when the Communists overtook southwest mainland China. The much larger group of Chinese in Táiwān includes both KMT emigrants and Chinese already present on the island before the Communists overtook mainland China (and the descendants of the people in those two categories). These two, non-mainland Chinese societies share 5000 years of relatively undisrupted Chinese tradition. The Chinese Communist Party (CCP) on the mainland systematically disrupted the development of Chinese culture, with traumatic consequences, while the Táiwān Chinese and Northern Thailand KMT villagers (non-mainland populations) continued to emphasize and celebrate it. Hence, we can treat the collective memory of the Táiwān Chinese and Northern Thailand KMT villagers as 100% of the possible collective memory for traditional Chinese culture, a normalized baseline against which we can compare the collective memory of the mainland Chinese.

To facilitate our analysis, we will divide the time frame of Chinese collective memory into five periods: antiquity (up until 1910), the pretrauma period (1911–1949), the trauma period (1950–1976), the posttrauma period (1977–1989), and contemporary China (1990 onward). The period of antiquity begins at an unspecified time in the distant past and ends with the end of the Qīng Dynasty. The pretrauma period is one of turmoil that includes the Second Sino-Japanese War (1937–1945) and the Chinese Civil War (1945–1949); it ends with the Communist takeover under Máo Zédōng. The trauma period is one of disruptive social programs led by Máo that include the Great Leap Forward and, especially, the Cultural Revolution; it ends with Máo's death. The posttrauma period begins with political maneuvering that resolves into a moderate liberalization under Dèng Xiǎopíng; it ends with the protests at Tiananmen Square (1989). The contemporary Chinese period begins with the presidency of Jiāng Zémín and is characterized by the rapid economic development and very gradual liberalization that continues today.

Following our memory consolidation model described in preceding chapters, we propose that the Communist takeover and subsequent Communist social programs (1950–1976) interrupted declarative memory consolidation and therefore reduced or erased mainland Chinese declarative memory of the pretrauma period (1911–1949), with relative sparing of older memory (before 1910). It is important to emphasize that we gauge memory loss by comparing mainland with non-mainland collective memory, not by measuring absolute amounts of collective memory (which is infeasible). As with individual memory (see chapter 2), there is gradual memory loss with time in collective memory also, so that the non-mainland Chinese obviously remember less about the ancient past than about more recent times. But we gauge the relative memory loss of mainland Chinese against this background of non-mainland memory loss. Thus, another way to summarize our general finding is that mainland and non-mainland Chinese have similar levels of remote collective memory, but that mainland Chinese have reduced recent collective memory compared with that of non-mainland Chinese.

Crucial for our analysis of collective retrograde amnesia is the "relater" (relationality) element of our model. The social analogue of the relater is the set of individuals in a society who act as the primary "memory makers" for that society. A society can have many groups, each with their own relaters who can interact between groups, forming "the relater" for the larger society. In China during the pretrauma period, intellectuals, Communists, and Nationalists, together comprising the relater in the larger Chinese society, debated the advantages and disadvantages of traditional Chinese culture (as it was understood in the 1800s and early 1900s) compared with Western cultures. In 1949, with the establishment of the People's Republic of China (PRC), the CCP effectively won the debate on the Chinese mainland and established itself as the exclusive relater in mainland Chinese society. From roughly 1950 and into the late 1970s, the CCP attempted to replace traditional Chinese culture—and collective memory—with Marxism and state-approved memories.

By comparing mainland with non-mainland collective memory, we will try to gauge the relative amount of mainland Chinese collective memory that existed during the posttrauma period, as defined earlier. As a result of the replacement of traditional intellectuals with Communists during the trauma period (1950-1976), our research indicates that the collective memory of the mainland Chinese during the posttrauma period

(1977–1989) was robust for the ancient past but was diminished significantly for the pretrauma period (1911–1949). As exemplified primarily in the domains of religion and literature, we see that very old memories—the ones that had consolidated generations before—were durable. However, some newer memories that we can reasonably assume would have been consolidated were lost because of the Communists' traumatic social disruptions. This chapter introduces the societies of interest and provides the necessary historical background. Chapter 10 describes the persistence of consolidated religious schemata for Confucianism, Daoism, and Buddhism. To highlight memory loss from the pretrauma period, chapter 11 examines a group of relaters who surfaced in mainland China beginning in the late 1980s: the authors of the so-called "Thinking Generation." These authors have striven to reclaim the trajectory that Chinese culture had assumed prior to the revolution and have also enabled contemporary mainland Chinese to retrieve collective memories from the pretrauma period.

The Social Hippocampus

As we discussed in chapter 2, lesioning (physically destroying) the hippocampus disrupts the consolidation of individual declarative memories, thereby resulting in individual retrograde amnesia. Although retrograde amnesia can refer to the loss of any or all memory from before the time of the amnesia-causing insult, we will use the term "retrograde amnesia" to refer to that specific pattern of memory loss that follows lesioning of the hippocampus in individuals: loss of memory for recent events but sparing of memory for events that occurred longer ago (remote events). As we discussed in chapter 6, the main function of the hippocampus is to establish the systems of relationships between short-term (and some long-term) memory items that are used to form and gradually reshape long-term memory. As the relationality element (or relater), the hippocampus plays the central role in the consolidation process, and its removal prevents further consolidation (see chapters 2 and 6 for details).

We propose an analogy between the hippocampus of an individual and the social hippocampus of a society. Any group can have a social hippocampus, so a larger society composed of many groups will have many "social hippocampi." They are made up of the "opinion leaders" or "meaning makers" or "memory makers" of a society—its religious specialists, politicians, prominent government officials, scholars, journalists, and publishers—who select memory items from the recent past, relate recent and more remote items, and prepare them for consolidation into stable collective memories. Social hippocampi are not only the people but also the institutions that have been set up to support these meaning-making activities. They function in a society in the same way that the hippocampus does in a human brain: Social hippocampi establish the relationships between memory items that give them meaning and place new (unconsolidated) and old (already consolidated) items into an organization that supports the development and maintenance of consolidated, stable collective memory.

Although we focus here on the social hippocampus or collective relater, we also recognize collective analogues of the buffer, generalizer, and entity elements of our three-in-one consolidation model (these analogies are described in detail in part II). For our collective retrograde amnesia study, the entities in question are the three peoples we compare: the mainland Chinese and the populations of Táiwān and Northern Thailand. In individuals, the buffer and generalizer are both mediated by the largest part of the forebrain (see chapter 2), and labile and stable memories are stored there. By analogy, the collective buffer and generalizer are mediated by the society at large, and they and their memory structures are the substrate for both labile and stable collective memory. Thus, declarative labile and stable collective memory is what individuals can collectively declare that they remember through interactions with each other and by reference to short-term and long-term memory structures such as newspapers, the Internet, other media, books, monuments, memorials, museums, and so forth. As in individual declarative memory formation, the collective relater plays a pivotal role as it interrelates labile (and some stable) items from the collective buffer and presents those to the collective generalizer for consolidation into collective stable memory. Our analogy with the neurologic hippocampus emphasizes the central importance of the social hippocampus in the process of collective memory consolidation.

The main difference between the neurologic hippocampus and the social hippocampus is that the former, once removed, never grows back, whereas the latter is easily replaced. More often than not, a powerful group that destroys an existing social hippocampus then replaces it with one of its own. Removal of the hippocampus in an individual causes both retrograde (described earlier) and anterograde amnesia. The latter is the inability to remember events that occur after the insult that causes the amnesia. The removal of a social hippocampus should produce the collective analogues of retrograde and anterograde amnesia. However, if a new social hippocampus takes the place of the one that was destroyed, then consolidation should continue and there should be no anterograde amnesia, although we would expect a selective retrograde amnesia for the specific items that were being consolidated by the social hippocampus that was destroyed. This situation occurred as a result of the Cultural Revolution.

During the Cultural Revolution (CR), the Communists under Máo Zédōng destroyed the social hippocampus (SH) operating at the time (the pre-CR SH) and replaced it with one of their own (the post-CR SH). The collective memories that were being consolidated by the pre-CR SH were lost when the pre-CR SH was destroyed, and this produced the collective retrograde amnesia that we will describe in chapters 10 and 11. An antero-grade amnesia would also have occurred if no post-CR SH had been put in place, but collective memory consolidation continued in Communist China after the CR through the post-CR SH set up by the CCP. The post-CR SH, operating under the aegis of the CCP, had no interest in consolidating the potential memory items that would have been of interest to the pre-CR SH. Because the CCP installed its own post-CR SH, we cannot consider a collective anterograde amnesia in the case of Communist China. Instead, we keep our focus trained on collective retrograde amnesia.

The Cultural Revolution as a Social Trauma

The Communist movements that took place from 1950 to the late 1970s, especially the Cultural Revolution from 1966 to 1976, are an example of a social trauma that causes collective retrograde amnesia. The Cultural Revolution specifically targeted the existing "meaning makers" or the "social hippocampus" of mainland Chinese society. Of course, the Com-munists installed their own social hippocampus, which attempted to estab-lish its own version of Chinese history and culture. This "hippocampal transplant" (a term from which we will henceforth refrain to avoid belabor-ing the analogy) will not complicate our analysis. The targets of the Cul-tural Revolution were so clear, and its intentions were so obvious, that we can essentially ignore the post-Communist culture and focus instead on the ways in which memory of the traditional, pre-Communist Chinese culture was affected by the Cultural Revolution.

Communism was introduced into China in the first decade of the twen-tieth century, when colonial control by industrialized countries finally destabilized the 2000-year-old Chinese monarchy. Facing the impotence of the Qīng Dynasty (1644–1910), educated Chinese began to seek ways to save China from complete colonization. In light of the military and economic dominance of Western nations, these Chinese also felt com-pelled to evaluate and critically re-examine traditional Chinese culture. A great debate sprang up between radical and conservative intellectuals in the early 1900s. Led by Chén Dúxiù (陈独秀, 1879–1942), who later became the first president of the CCP, the radicals wanted to completely abandon traditional culture and uncritically introduce Western scientific knowledge and democratic social systems. The conservatives, exemplified by writers of the Xuéhéng school (学衡) and the Peking school, tried to reconcile essential elements of traditional Chinese culture with advanced scientific knowledge from the West and to save China without losing Chinese identity (Stuckey, 2005). The debate between Chinese radicals and conservatives can be seen as an example of the "contest and negotiation" (Wertsch, 2002) that accompanies the relationality process that is central to collective memory consolidation. This process of relationality would ultimately be disrupted by the CCP.

In the early 1900s, "the radicals" included both future Nationalists and future Communists. They disagreed on many issues, including which form of Western thought to introduce into China. In 1911, the hereditary imperial government was overthrown and replaced by the Nationalist Republic of China led by the KMT. Infighting, civil war, invasion by Japan, and the Second World War burdened the next four decades, and the turmoil allowed the Communist radicals to gain increasing numbers of supporters. Exploiting the growing Chinese interest in Western thought, early Communists such as Lǐ Dàzhāo (李大钊, 1888–1927) and Chén Dúxiù propagandized Communism and called for a social revolution, the abandonment of traditional Chinese culture, and the transformation of China into a "modern" nation-state.

In Shànghǎi in 1921, the CCP was founded with the ultimate goal of transforming China into a Socialist, and then a Communist, country. The CCP rapidly gained power. By 1949, the CCP had succeeded in confining the KMT to the island of Táiwān, and in establishing the PRC on the mainland. As part of the CCP's effort to spread Communism in China, conservative intellectuals and their ideas about traditional Chinese culture, religion, and governance were targeted by the Communists for severe repression, if not outright removal, and the CCP attempted to systematically implant Marxism in China. For example, in the PRC's first constitution (1954), Marxism was considered the supreme guideline for government policies and Communism the ultimate goal for China. A decade later, to legitimize fully Marxism and Communism at the grassroots level, the CCP attempted to remove all non-Communist elements from Chinese daily life. This series of political movements reached its crescendo during the Cultural Revolution (1966–1976).

The Cultural Revolution was as much a political battle between Chairman Máo Zédōng (毛泽东, 1893–1976) and his political enemies (for example, Liú Shàoqí [刘少奇, 1898–1969] and Dèng Xiǎopíng [邓小平, 1904–1997]) as it was a cultural battle between Marxism and traditional Chinese culture. Máo wanted to assert his political dominance, and forcing China to undergo the Cultural Revolution was part of his political agenda. But Máo also wanted to wash the Chinese mind of all non-Communist ideas, and he was ruthless in pursuit of this goal. Organized religion as well as private rituals were forbidden, temples and monasteries were closed, ceremonies were discontinued, and spiritual leaders were forced to return to secular life. Non-Communists were harassed and prevented from communicating with each other and from promulgating ideas with which the state did not agree. Unapproved books were banned, independent media outlets were shut down, and countless historical artifacts were destroyed. Specialists who conserved and disseminated cultural knowledge—college faculty members, scholars, writers, artists, newspaper editors, publishers, broadcasters, and others—were either "reeducated" until they completely capitulated to the CCP or they were jailed or even killed. Consequently, common Chinese lost access to historical records and non-Communist works (e.g., Chinese classical philosophy, literature, and all non-Communist foreign theories). Finally, 5000 years of Chinese history was rewritten according to the class-struggle theory of Marxism.

The Cultural Revolution had truly traumatic effects on Chinese collective memory. In terms of our memory consolidation model, it purged labile non-Communist items from the collective memory buffer, and it simultaneously interrupted the functioning of the collective relater. By "interrupted" we mean that the individuals and organizations that made connections among and between newer, short-term memory items and older, long-term memories were suppressed or eliminated. With their actions restricted and their memory formation structures removed, the traditional Chinese mainland society, because of its "lesioned" social hippocampus, could not continue its ongoing process of collective memory consolidation. The only collective memories consolidated in mainland China during this period were those sanctioned by the CCP: They were carried by official media into the collective buffer, related by party officials and state-approved university faculty, and condensed down into a version of history and current events that was consistent with the party line.

The loss of traditional relaters in Chinese society during the Cultural Revolution had consequences that extended well beyond the period of trauma. At the end of the Cultural Revolution, the most severe restrictions were removed, records were made available, and specialists were allowed to practice and teach again. Yet younger people, those born after the establishment of Communist China, had difficulty integrating and consolidating the remnants of traditional culture to which they were newly exposed (see chapter 11). They did not have enough knowledge of traditional culture to establish meaningful relationships between it and newer memory items. As we will see, it took most sectors of Chinese society about 10 years to establish such relationships, during a later period of liberal reform in the late 1980s. Our exploration of how the Cultural Revolution interrupted memory consolidation for pretrauma, non-Communist memory items will provide evidence for the existence of collective retrograde amnesia between 1976 (when the Cultural Revolution officially ended) and the late 1980s (when the interrupted memory consolidation process could be resumed).

The Cultural Revolution differentially influenced the collective memory of older and younger generations on the Chinese mainland because of unequal access to items that were in the process of being consolidated during the pretrauma period. We are particularly interested in the collective memory of the generation born in the 1950s, known in China as the "Thinking Generation." As a product of the CCP's Communist education, they did not have direct experience with, and were denied access to, memory items that otherwise would have been current during the pretrauma period. Because the Thinking Generation's memory consolidation process for the pretrauma period was disrupted, focusing on this slice of mainland Chinese society will provide the best evidence for memory loss from the pretrauma period.

Although the cohort born before the Thinking Generation lived through the pretrauma period and was exposed to memory items that were current during that period, they were unable to consolidate those items because of the disruption of the consolidation process that took place during the subsequent trauma period. That made it impossible for the preceding generation to pass recent memories on to the Thinking Generation. In contrast, the preceding generation was able to pass on memories that had already been consolidated. The result is a loss of recent collective memory but a relative sparing of remote collective memory, as exhibited best by the Thinking Generation. As we shall see in chapter 11, many collective memory items from the pretrauma period that had been lost to the Thinking Generation were recovered, beginning in the late 1980s, both because the CCP government had slowly adopted a more friendly attitude toward traditional culture and because mainland Chinese began to communicate with the international Chinese diaspora, which still possessed those memories. The end result is that the generations following the Thinking Generation were able to resume, at least to some extent, the previously interrupted course of collective memory consolidation that continues to this day.

The Thinking Generation and Their Literary Achievements

With little memory of pre-Communist China, the young people born in the early days of the PRC (1949–1959) were the least "corrupted" by the old society and therefore the most promising subjects for socialist transformation. In their teens during the Cultural Revolution (1966–1976), they were mobilized and empowered by Chairman Máo to beat down, literally and figuratively, the remnant feudal and capitalist elements in Chinese society. Squandering what little adolescent academic preparation they had, these youths devoted themselves to the mass struggle with passion. As a result, when social order was temporarily restored in the late 1960s, the government faced the problem of what to do with these premature revolutionaries, who could neither return to school (because a younger generation had to be educated) nor work in factories, as industrial production had been intermittently suspended since 1965.

The government finally decided to relocate the youths and have them work with peasants in the countryside, claiming that the young would help China bridge the "three disparities"—between town and country, between mental and manual labor, and between industry and agriculture—that still impeded socialist transformation. Facing the hardships of life in remote rural areas without knowing their future, many urban youths felt abandoned and betrayed by the leaders and the system for which they had fought. When they were legally allowed to return home to their cities and towns, most had difficulty finding jobs there. These practical difficulties stimulated some of them to produce "Scar Literature" (伤痕文学) in the late 1970s, criticizing the Communist movements and revealing the resultant suffering (Siu & Stern, 1983). It was that propensity for soul-searching that gave this cohort the name "Thinking Generation."

The literary achievements of the Thinking Generation were nourished by both Chinese and Western literature. At the end of the Cultural Revolution, the CCP's control of literature weakened, and people regained access both to Chinese and foreign literary works. In particular, a handful of widely circulated journals—such as *Journal of Foreign Literature in Translation* (译林), *Research in Foreign Literature* (外国文学研究), *Contemporary Foreign*

Literature (当代外国文学), and *Foreign Literature* (外国文学)—were published, providing the opportunity for Chinese readers to learn about previously prohibited Western literary theories (Ying, 1992). These new journals quickly acquired prestige and a large readership, stimulating the Chinese to rethink the Communist ideal of modern China and the traditional Chinese culture that had been denied or otherwise disrupted by the Cultural Revolution.

As more Chinese engaged in reevaluation in the mid-1980s, a serious debate erupted concerning comparisons between traditional Chinese culture and modern Chinese society and between Chinese and Western cultures. It was not unlike the debate that roiled in the early decades of the century, but after the miserable disaster that was the systematic abandonment of traditional Chinese culture and the uncritical use of Marxism since the early 1950s, many Chinese had regained an enthusiasm for their traditional culture. Among other things, this debate helped many Chinese realize that Western cultures alone could not save China. At the same time, the CCP government, led by President Dèng Xiǎopíng, launched the Movement for Reform and Opening-Up (改革开放) with the ultimate aim of economic recovery from 30 years of disruptive Communist social movements. Accordingly, in the late 1980s, the CCP government for the first time embraced the market economy that it once denounced as exploitative.

As a result of the increased liberalism that accompanied the mid- and late-1980s era of Reform and Opening-Up, the Chinese began to cultivate a "Cultural Fever" in which they embraced traditional Chinese culture and built a new national/cultural identity (Ying, 1992). "Cultural Fever" (文学热) stimulated yet other literary figures of the Thinking Generation to rethink their Chinese cultural heritage. Dissatisfied with the increasingly formulaic "Scar Literature," these writers called on a deep cultural awareness, or in their words a "cultural root," in order to find new (for them) forms of expression. In 1985, Hán Shàogōng (韩少功, 1953–) published the foundational text of this new movement in his essay, "The 'Roots' of Literature" (文学的'根'). Propelled by Hán and others, "Root-Seeking Literature" (寻根文学) emerged as a forceful cultural critique of the aesthetic and ideological dogma of official Communist literature.

The majority of the root-seeking writers, including Hán Shàogōng, Ah Chéng (阿城, 1949–), Zhāng Chéngzhì (张承志, 1948–), Wáng Anyì (王安忆, 1954–), Zhèng Yì (郑义, 1943–), Lǐ Hángyù (李杭育, 1957–), and Lǐ Ruì (李锐, 1950–), were born in the 1940s and 1950s and grew up and were educated "under the red flag." The majority of the root-seekers lacked decent formal education because of governmental failure to promote non-political education during the Cultural Revolution. Neither had they any opportunity for systematic self-education, due to the CCP's suppression of non-Communist resources. The only background they could rely on to cultivate their thinking was their life experience, in the remote country-side, as urban youths who were sent to be reeducated by peasants. This experience provided them a lived cultural tradition as an alternative to the denied or distorted urban experience of their childhood. They responded by making the countryside a locus for exhibiting a complex of opposing values including rural ethics versus urban mores and lyrical escapade versus political engagement (Wang & Chi, 2000).

Such cultural awareness made the root-seeking writers the heirs to a large number of traditional Chinese writers who advocated "returning to the ancient." However, unlike traditional writers, root-seekers in the 1980s were rarely concerned with the Confucian ideals of "poetry as education" (诗教) or "music as ritual propriety" (礼乐). Rather, they celebrated the unpolished purity of the lives of people in the remote countryside and paid homage to nature, as is characteristic of Daoism and Buddhism (Li Qingxi, 2000, pp. 115, 122). In consideration of the Thinking Generation's literary achievement and cultural awareness, we focus on the Thinking Generation's memory of the three major Chinese religions (Confucianism, Daoism, and Buddhism), as expressed in their literary works (chapter 10), and of Chinese literature from the pretrauma period (chapter 11). Our examination of Chinese religion and literature will give us a view of the recent and remote collective memory, respectively, of the Thinking Generation.

The Cultural Revolution Targeted Chinese Religion

The state of religious practice in mainland China has been in flux through-out the history of the PRC. Despite the materialistic and atheistic Marxist theory of religion, the CCP formally promulgated a policy of religious freedom in three successive constitutions, in 1954, 1975, and 1982 (Dillon & Minority Rights Group, 2001). This suggests that, at least in principle, the CCP's aim was to control religions, rather than completely eliminate them. To implement that control, the CCP sponsored the Religious Affairs Bureau (RAB) in 1954 to supervise religious activities. Temples, churches, and mosques were required to register with the RAB (now known as the State Administration for Religious Affairs [SARA]). During the rela-tively liberal periods of 1950–1956, 1960–1965, and after 1979, these institutions were given a limited degree of autonomy. However, in radical periods such as the Anti-Rightist Movement (1957) and the Great Leap Forward (1958), religions were denounced as one of the pernicious "Four Olds" (old ideas, old culture, old customs, and old habits). The Cultural Revolution further attempted to eliminate all religions and replace them with Marxism, which was supposed to satisfy people's spiritual need for purification and rebirth with socialist substitutes such as public confessions and the Cult of Chairman Máo (Welch, 1972).

Above all, the CCP demanded that its citizens pledge their ultimate allegiance to the state and the party, not to a deity or religious leader. Under the Cultural Revolution, the three major Chinese religions (Confucianism, Daoism, and Buddhism) suffered different fates, owing to their different origins and relations to government power. Before 1911, Confucianism was the source of imperial legitimacy; Daoism was the most popular folk religion without much involvement with government ideology; and Buddhism, imported from India and indigenized in China, provided a vital connection between China and other Buddhist countries. Consequently, the CCP appropriated Confucian ideals for governing while suppressing Confucian practices. Folk Daoist practices survived at the grassroots, even though the CCP removed their religious interpretive devices (e.g., professional practitioners). Buddhism was domestically suppressed even as the CCP government appropriated its legitimizing effect in its international relations with Buddhist countries in mainland Southeast Asia.

These three different outcomes imply that the results of the CCP's efforts at replacing all religions with Marxism were ambiguous. In other words, it would not be fair to say that all Chinese collective memory for religious ideas and practices was wiped out. As professor of philosophy and Chinese religions, Wing-Tsit Chan (1901–1994), commented at the time: "Communist triumph in China seems on the surface to forecast the termination of religion in China. Viewed in the historical perspective and in the patterns of Chinese religious thought, however, Chinese religion has outlived many political systems and ideologies" (W. Chan, 1978). We found that Professor Chan's prediction concerning the survival of Chinese religions was, to a large extent, valid. However, the outcome of the interaction between Communism and China's three major religions provides crucial evidence to support our theory of collective retrograde amnesia. Specifically, we found evidence for the persistence of memory of an ancient (remote) past vis-à-vis reduced memory of the immediate (recent) pretrauma period. Our research suggests that only certain facets of religious memory were disrupted, such as the understanding of the specific religious meaning of certain surviving rituals, whereas the basic tenets of the three major religions—Confucianism, Daoism, and Buddhism—remain ingrained in mainland Chinese society. We detail these findings in chapter 10.

The Cultural Revolution Targeted Chinese Literature

In tandem with their attempt at replacing religion with Marxism, the CCP tried to replace a diverse, modern Chinese literature with a monolithic proletarian literature that they used mainly as a tool with which to extol the virtues of the ideal Communists: peasants, workers, and soldiers. Modern Chinese literature had begun with the vernacular literature movement in 1917, when Hú Shì (胡适, 1891–1962) and Chén Dúxiù in turn published two significant articles on literary reform in *New Youth*, the leading journal of Chinese intellectuals in the 1910s and 1920s. They advocated writing in the modern vernacular, rather than in classical Chinese. This new style was rapidly adopted over the next few years and was to become more firmly established after the May Fourth Movement, which was an intensification of what was known as the New Culture Movement. The May Fourth Movement started as a demonstration on May 4, 1919, against the Běijīng government, in protest of its decision to cede the province of Shāndōng (山东) to Japan. This event invigorated and strengthened the cultural movement led by early Chinese Communists who wanted to reform Chinese traditional culture by adopting a more Western attitude.

Catalyzed first by international interest and then by the May Fourth Movement, many Chinese writers absorbed foreign literary theories, especially from Europe, Japan, and Russia. Chinese studies specialist Lee Leo Ou-Fan has traced the impact, for instance, of the European "romantic" tempo on a number of modern Chinese men of letters (Lee, 1973). And sinologist Bonnie McDougall (1977) observes in her widely acclaimed paper, "The Impact of Western Literary Trends," how Western literature enjoyed enormous popularity from its introduction to China in the early 1910s until the outbreak of the Second Sino-Japanese War (1937–1945). Beginning in the late nineteenth century, when an increasing number of Chinese students went to study in Japan, Japanized Western literary theories and Japanese writers such as Mori Ōgai (1862–1922) and Ntsume (Kinnosuke) Sōseki (1867–1916) had influenced Chinese writers. For example, East Asia literary scholar Ching-mao Cheng discusses how Guō Mòruò (郭沫若, 1892–1978), a well-known Chinese writer who had been educated in Japan in the 1920s, contributed to the theorization of modern Chinese poetry (C. Cheng, 1977). Similarly, Russian literature gained great popularity among modern Chinese revolutionary writers starting in the 1920s, as the CCP promoted Russianized Marxist ideas and social systems as an ideal for Chinese Communists. Chinese scholar Mau-Sang Ng's (1948–1994) major contribution to literary studies was his research showing how the supreme Russian masters like Tolstoy, Dostoyevsky, Turgenev, Chekhov, and Gorky had shaped the minds of key May Fourth writers (Ng, 1988).

This diversified development of modern Chinese literature thrived for no more than two decades before it was smothered by anti-Japanese sentiment after the Japanese bombing of Shànghǎi in 1937. Most writers fled to the hinterlands, where Máo Zédōng's dictum of literature as an instrument for politics eventually dominated creative writing. In 1942, Máo announced in his "Talks at the Yán'ān Forum on Art and Literature" that "Literature and art are subordinate to politics"; he aspired to channel all literary activity and artistic thought toward one common, Communist purpose, and to establish revolutionary and proletarian literature and art as the only "good" kinds (McDougall & Máo, 1943/1980, pp. 75–78). As the CCP intensified its influence throughout China in the ensuing years, Máo's doctrine gained popularity and effectively narrowed modern Chinese literature into an increasingly dogmatic and mechanical application of an extremely determinist and anti-humanistic ideology (Duke, 1985).

This subjugation of literature reached its peak during the Cultural Revolution, a period in which the CCP controlled all publication. The extent of this control is starkly illustrated by statistics compiled by literary historians Helen Siu and Zelda Stern. "In 1960," they write, "1,300 official periodicals were published in China. In 1966, at the beginning of the Cultural Revolution, the number was cut to 648, and by 1973, this number had been further reduced to about 50" (Siu & Stern, 1983, p. xlv). During this period, all foreign non-Communist literature and traditional Chinese literature was barred from publication, distribution, and library circulation. This included the modern Chinese classics produced by the May Fourth writers in the 1920s and 1930s. The only works widely available were proletarian literature, political tracts, technical manuals, a few novels, and works by Marx, Lenin, and Máo. All literary works expressing emotions antipathic to the revolution and non-Marxist understanding of the times—known today as "Underground Literature"—were forbidden. Worse, CCP officials routinely threatened writers with public humiliation, imprisonment, exile, or even execution, forcing them to select their fate from a list of alternatives that was limited to obedience, silence, or death. As a result, "by 1977, Chinese literature had reached one of the lowest points of its entire three thousand year history" (Duke, 1985, p. 185).

After the Cultural Revolution, the CCP began to relax its stranglehold on Chinese literature. After the death of Máo in 1976, the Maoist distinction between "bourgeois" high literature and "mass" literature was no longer enforced by many CCP officials (Link, 1983; Kinkley, 1985). Many formerly denounced writers, such as Bā Jīn (巴金, 1904–2005) and Shěn Cóngwén (沈从文, 1902–1988), who had produced most of their works before 1950, were rehabilitated and allowed to write again. The writers of the Thinking Generation started their literary exploration in this liberated atmosphere. They published many of their works in *Today*, a leading representative of the unofficial journals that provided an informal forum for young writers and readers in Běijīng. These youths advocated the independence of literature from politics. In their own words, "literature should not be a merely political statement or let itself be used as a tool for political struggle, as had often been the case in the past. Literature should serve as a vehicle to express people's feelings, as a bridge between souls, and as a means of purifying people's natures" (Pan & Pan, 1985, p. 195). The journal *Today* was widely considered a leading literary outlet in which a youthful "underground literary circle" published non-proletarian works.

To summarize, the traumatization of modern Chinese literature from 1950 until 1976 was the result of the monopoly of Communist proletarian literature and the hegemony of Máo's doctrines on culture and artistic expression. With their attacks on dissenting authors and ideas, the Communists interrupted the consolidation of non-Communist literature from the first half of the twentieth century. Chapter 11 will examine the Thinking Generation's memory (or, rather, its amnesia) for four well-known writers from the pretrauma period: Shěn Cóngwén, Zhōu Zuòrén (周作人, 1885–1967), Zhāng Àilíng (张爱玲, 1920–1995) and Hú Shì, who were suppressed by the CCP from 1950 until 1976 because their views on literature clashed with CCP doctrine. Despite severe suppression, and the resulting amnesia for pretrauma literary achievements, elements of traditional Chinese culture (the three religions in particular) survived and resurfaced in the mid-1980s in the new works produced by the Thinking Generation. Thus, our analysis of the Thinking Generation's collective memory will uncover evidence both for the loss of recent memory and the sparing of remote memory and will support our hypothesis that social trauma, specifically affecting the relaters in a society, results in a collective retrograde amnesia analogous to that observed after hippocampectomy in individuals.

Mainland Chinese versus Táiwān Chinese/Northern Thailand KMT Villagers

Finally, let us introduce the specific populations that will constitute the subjects of our comparison. Our goal here is to compare two groups that share the same Chinese cultural heritage but not the same recent history. As we have just explained, we will focus on religious practice and literary expression, because these two aspects of Chinese cultural life are interrelated and together span a wide swath of historical time. In a larger social context, our comparison will involve the Táiwānese, on the one hand, and the mainland Chinese on the other. In a smaller social context, and one that will allow finer distinctions to be made, our comparison will involve Guómíndǎng (Chinese Nationalist Party, or KMT) villagers in Northern Thailand, on the one hand, and on the other hand the people of Héshùn Township (和顺乡) in southwestern China, from which region most of the KMT emigrated to Thailand. The comparisons focusing on literary expression mainly involve the Thinking Generation on the mainland versus their counterpart in Táiwān, and the comparisons focusing on religious practice involve the Héshùn people versus the KMT villagers and the Táiwānese more broadly.

Most of the data on Héshùn and Táiwān came from historical records, memoirs, genealogies, and government archives on Chinese history. Héshùn Township has the biggest township library in China, containing extensive collections on its own history, and Héshùn has become a favorite place for anthropologists and historians to study oral history and social memory. Much of the data on the Northern Thailand KMT villagers came from fieldwork reports and monographs by anthropologists Chāng Wen-Chin, Yáng Huì, and Duàn Yǐng, who began studying these villagers' identity construction and historical memory in the 1990s (W. Chāng, 1999, 2001, 2002, 2003, 2004, 2006; Duàn, 2002; H. Yáng, 2007–2008). Comparison on specific points of religious practice between these groups will allow both general and more specific features to be brought out.

The validity of the proposed comparison is based on the shared Hàn Chinese origin of most Táiwānese, the KMT sympathizers who fled to Thailand, and the Héshùn communities they left behind. Hàn is the majority ethnicity in mainland China. The Héshùn are Hàn Chinese who emigrated around the sixteenth century from Nánjīng, the current capital of Jiāngsū Province in east-central China, to southwestern China. Hàn Chinese migrated to Táiwān from eastern and southeastern China as early as the Táng Dynasty (ninth century), and the island had been part of the central Chinese imperial administration since the late Míng Dynasty (1368–1683).

The KMT government and the Táiwānese populace are predominantly Hàn. When the Chinese Communists took over mainland China in 1949, the KMT government, which had been in power since 1912, fled to Táiwān with its sympathizers. However, two KMT armies in Yúnnán Province (mostly in Héshùn and neighboring towns in southwestern China), the Third and Fifth Armies led by Generals Lǐ and Duàn, respectively, were blocked off by the People's Liberation Army and could not retreat to Táiwān. They fled to Burma in 1952, hoping someday to fight their way back to mainland China with support from the United States and the KMT government in Táiwān. The planned counterattack and illegal cross-border drug trafficking annoyed the Burmese government, which in the 1970s finally forced them to move southward into Northern Thailand. There they successfully defeated Communist forces for the Thai government, which granted them legal status in Thailand and allowed them to set up 67 villages in Chiang Mai and Chiang Rai (W. Chāng, 1999).

Obviously, these KMT villagers also experienced trauma during their migration, in the form of war, hunger, and disease. It is entirely possible that such experiences could diminish collective memory (e.g., Rossington & Whitehead, 2007). However, scholars have found that the condition of being forcibly expelled from a homeland often encourages displaced peoples to work that much harder to maintain their collective memory and identity (R. Cohen, 1997; Malkki, 1995). Such was the case for the KMT refugees. What trauma they experienced did not eliminate their collective memory for Chinese traditional culture; instead, it caused them to preserve it. Disconnected from the old and familiar, the villagers clung more tightly to their traditions, enabling them to reconstruct their Chinese identity in a foreign country.

It is not an uncomplicated Chinese identity they claim for themselves, however. Politically, the KMT villagers identify themselves with "fatherland Táiwān," where their compatriots live, whereas Yúnnán remains their native (but broken) motherland. From the 1990s, major financial aid from Táiwān funded agricultural development, medical services, handicraft training, infrastructure construction, and the development of Chinese schools in these villages. Accordingly, villagers continue to pledge their loyalty to Táiwān, teaching their children songs and recognizing holidays associated with Táiwān. Going to Táiwān for many of the youth not only means seeking a better education and career development but is also a way to discard their refugee status in Thailand and to return to their political fatherland (W. Chāng, 1999).

Our comparison will reveal the contrasting attitudes toward Chinese traditional culture taken by the settled vis-à-vis the diasporic Chinese. Mainland Chinese during the Communist movements were forced by the government to consider traditional culture as a barrier to modernization. By contrast, as a strategy to antagonize Communism on the mainland, the KMT in Táiwān have purposefully preserved their traditional culture, and the KMT villagers in Northern Thailand have deliberately emphasized their Chinese identity when negotiating between the Thai, the Táiwānese, and the mainland. They are grateful to the Thai for allowing them to stay, and they identify politically with Táiwān, but they remain emotionally attached to mainland China.

Conclusion

The case study presented here in part III will examine the collective memory of four groups in Chinese society: first, the memory of Héshùn Chinese of the Thinking Generation for religious practices, and second, that of writers of the Thinking Generation over the entire Chinese mainland for literature. (Recall that Héshùn is ·on the Chinese mainland.) Throughout, we compare collective memory on the mainland with that, third, of Northern Thailand KMT villagers (who emigrated from Héshùn Township into Thailand in the 1950s) and, fourth, of Táiwān Chinese. Of course, all collectives forget to some extent. Obviously, the non-mainland populations have not remembered everything they possibly could have remembered over the course of their history, but we treat the collective memory of the latter two groups as a normalized baseline, as 100% memory, against which mainland Chinese memory will be compared. We find evidence both of memory retention and loss.

The next two chapters compare the collective memory of the mainland Chinese after the Cultural Revolution with that of these two Chinese diasporic populations. What we find is evidence for collective retrograde amnesia in mainland Chinese society, specifically among the Thinking Generation of Chinese who were children and young adults during the traumatic and disruptive decade of the Cultural Revolution. On one hand, despite the Cultural Revolution, the Thinking Generation retained collective memory for ancient items, such as the central tenets of major religions. On the other hand, due to the Cultural Revolution, the Thinking Generation failed to connect old memory items, such as Daoist rituals, with their religious significance, and it lost collective memory for new memory items such as literary figures from the pretrauma period of the 1920s and 1930s. We explore these findings in depth in chapters 10 and 11.

The main goal of this chapter is to show that the trauma experienced by the "relaters" (the relationship makers, the social hippocampus) of mainland Chinese society during the Cultural Revolution did not eliminate the ancient (remote in time) collective memory of the mainland Chinese. Specifically, this chapter offers two kinds of evidence for the consolidation of the three major Chinese religions in Chinese collective memory. First, we will examine how Confucianism lastingly influenced Communism, despite the Chinese Communist Party's stated goal of eradicating the teachings of the Greatest Master. Second, we will explore the canonical facets of Daoism and Buddhism that, again, persisted despite the disruptions of the Cultural Revolution. The nature of religion is beyond the focus of our book (for recent treatments, see Boyer, 2001; Whitehouse, 2004). For present purposes, we conceive of the collective memory of religion as a complex form of consolidated, long-term, generalized collective memory that is partly narrative, partly social framework, and partly moral schema. Throughout this chapter, we will point out that the Cultural Revolution did succeed in eliminating more recent religious memory, including memory of events that link current moral values with the religious past, yet the central tenets themselves remain, showing that the essences of Confucianism, Daoism, and Buddhism have become woven into the fabric of stable Chinese collective memory.

The Trauma's Effect on Confucianism

The oldest of China's major religions, Confucianism (儒教) is sometimes taken as a synonym for China. Its founder, Confucius (ca. 550–ca. 479 BCE), was a peripatetic tutor of princes, whose teachings were preserved in a number of texts, primarily in the *Analects* (论语). As an ethical system entailing both religious and political aspects, Confucianism played a vital role in holding together imperial Chinese society from the second century BCE to 1911.

Yet early Communists and other members of the radical Chinese intelligentsia in the 1910s took a hostile attitude toward Confucianism and attributed China's relative slowness to it. For them, saving China meant discarding Confucianism. In 1912, before the Chinese Communist Party (CCP) was formed, Communists launched the New Culture Movement (新文化运动) to challenge the legitimacy of Confucianism as the supreme guideline for the Chinese people. In 1919, they led the May Fourth Movement and started a series of attacks on Confucianism as the reservoir of China's backwardness relative to the West. After the CCP officially formed, it quickly intensified its influence in China, and it founded a series of revolutionary bases in northern and western China in the 1920s and 1930s. During this period, it tried to replace Confucianism with Marxism as the only state ideology. With the establishment of the People's Republic of China (PRC) in 1949, the Communists believed they had the legitimacy and ability to destroy finally Confucianism's moral, religious, and educational roots.

To destroy the moral basis of Confucianism largely meant to erase the Five Essential Relations, a central tenet of Confucianism that regulates five basic relationships: emperor–subject, father–son, husband–wife, elder–younger brothers, and elder–younger friends. The relationship between emperor and subject is the model for the other four, and together they provided the structure of imperial Chinese society from the court to the family. This hierarchy also established a patrilineal family structure for people descended from the same ancestors; within the family, ancestral tradition exerted an overwhelming force upon offspring, just as an emperor had absolute power over his subjects. However, in 1911, the Guómíndǎng (Chinese Nationalist Party, or KMT) overthrew the Qīng Dynasty, ending the 2000-year-old Chinese imperial and monarchical system. They challenged not just the government but also its pervasive underlying sociocultural justification, Confucianism. In fact, both the Chinese Nationalist Republic (1912–1949) and the PRC (since 1949) encouraged the Chinese to pursue social equality among classes and ethnicities and between the sexes (Meisner, 1999).

As a result of this social leveling, recognition of the Five Essential Relations diminished. Disregard for the Five Essential Relations resulted in the decline of the worship of ancestors, which is the religious basis of Confucianism. From the 1910s, the CCP condemned ancestor worship as backward and superstitious. It forced people to dismantle their ancestral temples and burn family genealogies, which disrupted the ritual and moral basis of the traditional patrilineal family structure. It also broke the continuity that Chinese families had with the Confucianism of their ancestors, although, as we will see in the next section, it did not erase Chinese collective memory for the basic tenets of Confucianism.

To legitimatize Marxism fully, the CCP strove to eliminate Confucianism as the knowledge framework for gentry and aristocrats and as the social framework for common Chinese. Before 1903, there was still a nationwide decennary examination on Confucian knowledge from the Four Books (*Analects*, 论语; *Mencius' Words*, 孟子; *Great Learning*, 大学; and *The Doctrine of the Mean*, 中庸) and Five Classics (*Poetry Classic*, 诗经; *Great Emperor's Words Classic*, 书经; *Rites Classic*, 礼经; *Changes Classic*, 易经; and *History*, 春秋). The imperial government selected new officials on the basis of this examination. The Four Books and Five Classics were taught as the bible of Confucianism at all levels of school throughout China and functioned, at least formally, as the ultimate guideline for the courts of all imperial dynasties from the second century BCE, as well as for Chinese society as a whole. In 1903 the Qīng court, under pressure from reformers seeking new scientific and technological knowledge from the West, abolished the examination. After 1903, the Four Books and Five Classics were no longer taught in schools, and during the extreme leftist periods since 1949, they have been replaced by the works of Communist thinkers such as Karl Marx. During the Cultural Revolution, Máo's works, especially the *Little Red Book*, played the same role that the Four Books and Five Classics had before 1903.

The Persistence of Memory for Confucianism in Government

Though the CCP tried to replace Confucianism with Marxism, many aspects of Confucianism survived the Communist movements because the CCP itself relied on them to advance its own agenda. Early in 1956, sinologist David Nivison observed that modern Chinese thinkers had a tendency to incorporate aspects of Confucian thought into the "new" ideas they were developing in government, military organization, artistic expression, and other areas—even including Communism itself (Nivison, 1956). Ironically, its conflict with the KMT may have caused the CCP to take an intentional, but covert, step in the direction of Confucianism. In the late 1920s, the KMT government suppressed the rival CCP and annihilated most of the Communist forces in southern China. This forced the CCP to flee on "the Long March" to the north and west, where the KMT's control was weak. On the march, Máo established a revolutionary base in the city of Yán'ān (延安, in Shǎnxī [陕西] Province of northern China), where refugee Communists from the south gradually converged.

To recruit new soldiers, the Communists conjured up a model of "the good Communist." CCP leader and theorist Liú Shàoqí (1898–1969), who in the 1950s became the second president of the PRC, applied neo-Confucianist philosophy (first proposed by Wáng Yángmíng [王阳明, 1472–1529] in the sixteenth century) to develop a set of theories and practices on how to be a good Communist. Using the neo-Confucian notion that one should "seek for the truth in actual things," Liú proposed that a person who learned the ability of "self-cultivation" could learn to be a good Communist. "Self-cultivation" is the ability to keep before one's consciousness one's faults and one's ideals; such a Communist has an attitude of "genuineness and sympathy" (*Analects*, 15.24) toward comrades. He is "able to love others and save others" (*Analects*, 4.3), meaning he never injures comrades for his own benefit but vigorously fights enemies of the revolution. Like a good Confucian student, a good Communist is further capable of that most prized of Confucian virtues, the ability to "watch himself when he is alone" (cf. the Confucian Classics *Great Learning* [大学], p. 6; or *The Doctrine of the Mean* [中庸], p. 1). Liú's appropriation of Confucianism illustrates that Chinese social frameworks are so Confucian that the Communists had to use Confucianism to convey Marxist ideas. Far from eliminating Confucianism, therefore, the Communists actively preserved it, although they did not identify it as such but (re)presented it under the guise of a new Communist philosophy.

The CCP's attitude toward Confucianism was and is ambiguous. On the one hand, it openly strives to eliminate Confucianism and replace it with Marxism, whereas on the other hand, it purposefully but covertly appropriates the same Confucian ideology that had propped up the Chinese state for more than 2000 years. This ambiguity extended all the way up to Máo himself. While denouncing Confucianism, Máo took on many of the attributes, and allowed himself many of the privileges, that were reserved for the ideal Confucian ruler. In his quasi-emperor's role, Máo led the CCP government to embody the Confucian ideal of bureaucratic domination of the state over civil society—according to Confucianism, the state should take comprehensive responsibility to enrich and educate the people. Máo's works were worshipped during the Cultural Revolution, attaining the same exalted status that the Four Books and Five Classics of Confucianism had previously enjoyed.

The ambiguous relationship between the CCP and Confucianism also extended down to ordinary soldiers. To legitimize the CCP-backed government, the CCP in the 1950s built commemorative museums and temples for anonymous heroes who fought for the establishment of the PRC. Furthermore, it instituted annual commemorative rituals to consolidate Chinese memory for these heroes by encouraging common people to worship them as they might previously have worshipped family ancestors. The epic stories of these heroes were recounted in the CCP-sponsored history books read by students at all grade levels, potentially transferring memory of individual family ancestors to the founders of the regime.

To control the populace more effectively, starting in 1949 the CCP also appropriated Confucian techniques for governing the masses. This is nowhere seen more forcefully than in the area of agriculture. To mobilize the people into collective agriculture, the CCP deftly utilized Confucian principles: obedience to authority, respect for hard work, appreciation of frugality, emphasis on harmonious relations, and working for mutual benefit rather than accentuating individual differences and striving for individual profit. Such comprehensive and clever cultural manipulation was rewarded: In 1954, the CCP managed to feed more than 600,000,000 people on less than 13,000 square kilometers of arable land (W. Zhāng, 1999).

The CCP's appropriation of the paternalistic and communalistic Confucian style of governance demonstrates that many aspects of Confucianism not only survived the CCP's suppression but also shaped the CCP government. What we take from this fact is that these aspects of Confucianism had been well consolidated in Chinese collective memory before the Communists took power, and they continued to influence the collective decisions and actions of the CCP and so of China more broadly. In accordance with this view, American sinologist Thomas Metzger (1977) claims that modern Chinese intellectuals inherited the basic goals and aspirations of the Confucian tradition; what they accepted from the West (Marxism included) was nothing more than new tools to implement the same ancient goals and aspirations. Although Metzger's claim ignores the natural give and take of encounters between cultures, it nevertheless demonstrates the influence of Confucianism in modern (Communist) China. What the CCP attempted to change, in the end, changed the CCP, because the canonical elements of Confucianism have been so well consolidated into Chinese collective memory.

It might seem as though Confucianism persisted in China despite the depredations of the CCP because it was (and is) "ingrained" in Chinese society as a form of implicit memory. If implicit collective memory actually exists, then Confucianism may be part of the Chinese version, but the collective memory we identify here is declarative. Thus, Liú Shàoqí explicitly decreed that a good Communist should have an attitude of "genuineness and sympathy." Although he failed to identify these virtues as Confucian (*Analects*, 15.24), he clearly articulated them, after which many Chinese people "knew that" they had to cultivate these virtues in order to be "good Communists." Thus, the use of Confucianism by the CCP was covert but explicit. Its assimilation by the populace in the Communist context was facilitated by the fact that it was consistent with already consolidated social frameworks. Although most of the Chinese populace did not recognize that these edicts were repackaged Confucianism, they "knew" them in a declarative sense, in the same way that many Americans know that "love is patient, love is kind" but do not recognize it as an excerpt from a Biblical verse (1 Corinthians 13:4). The CCP's appropriation of Confucianism substantiates the persistence of religion after the Cultural Revolution as a declarative form of Chinese collective memory.

To highlight the consolidation of Confucianism in Chinese societies, let us also examine how Confucianism shaped the KMT government in Táiwān, where no social trauma like the Cultural Revolution interfered with collective memory formation. The KMT government had controlled mainland China from 1912 until 1949, before fleeing to Táiwān. Despite initially disparaging Confucianism the KMT overtly rehabilitated it, in part to further distinguish itself from the Communists. The KMT founder, Sūn Zhōngshān (孙中山, 1866–1925, internationally known by his Cantonese name Sun Yat Sen), urged that Confucianism should be resurrected as a source of strength and support for the government, calling on soldiers' and people's loyalty to the nation. Under his successor Jiǎng Jièshí (蒋介石, 1887–1975, internationally known by his Cantonese name Chiang Kai Shek), the KMT government launched the New Life Movement in the 1920s to establish a harmonious society through Confucian ethics.

After retreating to Táiwān in 1949, the KMT government proudly claimed Confucianism as their emblem of a simultaneously modern and authentic Chinese government and culture, and they condemned the Communist mainland for losing its cultural roots. In Táiwān, the worship of ancestors is still widely practiced; as a salient example, a 77th generation direct descendant of Confucius lives in Táiwān today and keeps an excellent memory of the family genealogy. Confucius's birthday is celebrated as Teacher's Day, when Táiwān Chinese conduct special services at Confucian temples that include dances dating from the Míng Dynasty (1368–1683). By contrast, most mainland Chinese do not know Confucius's birthday. In addition, Táiwān Chinese continue to hold an annual Dragon Boat Festival to commemorate the death of the great Confucian statesman and poet Qū Yuán (屈原, ca. 340–278 BCE). On the mainland, the festival and the poetry of Qū Yuán were forbidden during the Cultural Revolution (Hoobler & Hoobler, 1993). However, like the CCP government on the mainland, the KMT government in Táiwān has also exploited the Confucian view of the bureaucratic domination of the state over civil society (A.Y.C. King, 1996). Since the 1980s, Confucian morality has been credited as the driving force behind Táiwān's great economic achievements, providing fertile soil for new growth and development of Confucianism in the era of globalization.

The Persistence in Collective Memory of the Confucian View of Human Nature

For root-seeking writers in the 1980s, religion was *the* cultural root of Chinese society; it expressed basic cultural values as well as artistic principles. The root-seekers saw literature as a particularly powerful vehicle for these fundamental religious elements. In the context of our model, religious principles had been generalized through the process of Chinese collective memory consolidation and persisted as part of the stable social frameworks that guided culture, art, and other aspects of Chinese life. This section observes how the Confucian view of human nature continued to exist in modern Chinese literature despite the Cultural Revolution.

The issue of human nature has been a central one for Chinese philosophers ever since it was addressed by Mencius (372–289 BCE), the second Confucian saint after Confucius. Mencius made two assumptions about human nature that have persisted throughout the 2000-odd years of traditional Chinese philosophy and literature. The first assumption, and the most commonly shared one, is that human nature is originally good. So, for instance, the opening line of the Confucian masterpiece *Three Character Classic* (三字经, the first textbook for kindergarten education in imperial China), reads "Man is born with a nature that is originally good" (人之初,性本善). As a corollary of the first assumption, the second claims that man's nature is corrigible. Since Mencius, these two assumptions are axiomatically taken as given in many areas of traditional Chinese literature and philosophy (Poa, 1992).

In imperial China, these two assumptions enabled Confucians to educate and regulate common Chinese using Confucian values, regardless of their class, race,.and previous individual deeds (or misdeeds). The goodness assumption presumes that all people are good-natured and that all should want to be part of a harmonious society regulated by Confucian moral values. The corrigibility assumption presumes that even people who have done wrong can be redeemed through proper education. These two assumptions enabled Confucians to fit everyone, regardless of background, into the Confucian moral framework, and they existed in generalized form as part of the social framework that guided the behavior of the Chinese common folk.

Ironically, these two assumptions were also compatible with the aspirations of the May Fourth intellectuals who wanted to "save China"—by eliminating Confucianism! A particularly good example of this comes from author Zhōu Zuòrén, one of the founders of modern Chinese literature. In his parallel readings of European "humanism" and Chinese assumptions of an essentially good human nature, Zhōu translated the Renaissance discovery of "Man" into Chinese-Confucianist conceptions of human nature, so as to propose his notion of the "Literature of Humanity" (人的文学). He claimed that literature stimulates and releases the bright side of human nature, just what was needed to transform the Chinese nation in the early 1920s.

In the late 1920s, the ongoing development of Communism in China converted the literary discussion of human nature into a political issue. In a move that owed as much to increasingly sophisticated sociological theories as to their sociopolitical biases, Marxists claimed that in a class-based society, one's human nature always bears the stamp of one's class. According to this view, members of the proletarian and of the capitalist classes actually possess different natures, the one good, the other bad (Poa, 1992). Meanwhile, rightist writers such as Liáng Shíqiū (梁实秋, 1903–1987) refuted the class-based theory and promoted the idea that literature is pan-human and free of class background. The debate persisted for over a decade until the right-wing writers gave out, under the pressure the CCP exerted from its stronghold in northwestern China. At the Yán'ān Forum on Art and Literature in 1942, Máo singled out the concept of "human nature" as one of the erroneous ideas that had been corrected: "in a class society, there is only human nature that bears the stamp of a class, and no human nature transcending classes" (McDougall & Máo, 1943/1980). From then and until the late 1970s, Máo's doctrine was the guideline for literary creation.

With the relative relaxation in censorship after the Cultural Revolution, writers began to reclaim the idea of an essentially good human nature. Much as their May Fourth predecessors had disparaged the "feudalistic ethical mode" of the imperialists, writers in the 1980s criticized the Gang of Four for trying to eliminate and distort the "sprout of goodness" while the writers championed their own efforts to nurture it.[1] So it is evident that, despite the best efforts of the most ruthless Communists, the Confucian ideal of a bivalent but essentially benevolent human nature persists in Chinese culture. In the context of our model, the Confucian ideal had been generalized by the process of Chinese collective memory consolidation and had become part of the stable social frameworks that supported Chinese common and cultural life.

The Trauma's Effect on Daoism

The Chinese philosophers Lǎozí (老子, sixth century BCE) and Zhuāngzǐ (庄子, fourth century BCE) proposed Daoism as a school of thought in the Spring and Autumn Period (770–476 BCE) and War Period (475–221 BCE) of pre-imperial China, respectively. Daoism as a religion with ritual practices, organizations, and temples came into being in the second century CE as a combination of this philosophy and of the folk cultures of southern China. The word Dào (道) literally means "road" and is figuratively used in Daoism to refer both to the origin of the world and the way to the origin. Followers of Daoism endeavor to achieve immortality by embodying the "Dào." Daoist priests preside over temple sacrifices and festivals, and common practices include martial arts, astrology, fēng shuǐ (characterization of places as either fortunate or unfortunate according to the flow of moral forces), and living a life of humility and compassion. Unlike Confucianism, Daoism became established and enmeshed in Chinese culture largely independently of government. Both its organized and unorganized components have shaped almost every aspect of Chinese culture. Chinese writer and thinker Lǔ Xùn, who was enthusiastically hailed throughout the twentieth century and actually praised and celebrated by the CCP, once claimed that "the whole basis of Chinese culture is Daoism" (Fù, 1984). Perhaps because they believed in the truth of that statement, the CCP adopted an extremely hostile attitude toward Daoism.

Daoist beliefs and rituals deeply influenced the daily lives of common Chinese people and imposed barriers to the establishment of a Marxist social order within Chinese society. Consequently, Communists attacked Daoism as "the opium of the masses." From 1958, Daoist priests were

1. The Gang of Four—Jiang Qing (Madame Mao), Zhang Chunqiao, Yao Wenyuan, and Wang Hongwen—was the CCP oligarchy in charge of much of the Cultural Revolution.

forced to return to secular life, and temples and monasteries were either destroyed or appropriated as command posts for collective farms. Most of the centuries-old bells and Daoist alchemy furnaces were melted in the Steel-Making Movement of 1957–1958 (Fù, 1984).

Daoist classic literature was forbidden outright during the Cultural Revolution, but this was not the only obstacle the CCP placed between the people and Daoist texts. Even though many Daoist texts are extant, they are unintelligible to common people because they are written in classical Chinese and complex characters, and the CCP's Language Reform promoted modern, vernacular Chinese and simplified characters. But the common people, in any case, were unlikely to delve deeply into Daoist texts. As do laypeople in most cultures, the Chinese laity focuses more on ritual performance than on underlying meaning. The preservation of religious meaning depended on Daoist priests, and without them there were no religious specialists to preside over rituals or provide authoritative interpretations of Daoist texts. The Communist crackdown thereby removed the agency of interpretation for rituals and other Daoist practices from the common people on the Chinese mainland (Dubois, 2004).

A few examples will illustrate the point. Daoism gives "spirit" to the traditional Chinese arts. During the Six Dynasties (from the 3rd to 6th centuries), Daoist calligraphers adopted two ancient Chinese calligraphic forms of talismanic writing, Lìshū (隶书) and Zhuànshū (篆书). Daoists used this talismanic writing to communicate with supernatural beings, and many Chinese people hung a piece of Daoist talismanic writing on the main doors of their houses to protect them from evil spirits (D. Táng, 2001). The priests encouraged this practice and explained its meaning to the people. Later, during the Cultural Revolution, all talismanic writings were forbidden. Because professional Daoists were forced to go back to secular life, common people lost access to this writing and its religious meaning.

Another example concerns traditional Daoist architecture, with its emphasis on symbolism and geomancy. The two most important symbols, the tortoise and the snake, represent longevity and the Great Warrior, respectively. Accordingly, a Daoist architect prefers the groundsill of a building to be built on a tortoise-shaped base and the building to be backed by a "black dragon," usually a long (dragon-shaped) mountain (Hartz, 1993). A salient example of this kind of architecture is Kūnmíng (昆明), the capital city of Yúnnán Province. Designed by a Daoist during the Yuán Dynasty (1271–1368), it is located in front of a long mountain, which the local people call Snake Mountain. The city's hexagonal shape and six gates resemble a tortoise shell with its six outlets, one for each of the four legs, the head, and the tail. During the Communist movements, however, Daoist geomancy was denounced as a feudal superstition, and the Communists did their best to suppress transmission of this architectural tradition. Since the Cultural Revolution, only a few people know why the Kūnmíng city plan takes the shape of a tortoise; many are not even aware that it has this tortoise shape.[2]

Something similar occurred at Tài Mountain (泰山) and Wǔdāng Mountain (武当山), which are 1000-year-old Daoist sacred places. Since the 1980s, non-Chinese tourists have flocked to these two destinations to view their many fine examples of Daoist architecture. The religious meaning attached to these buildings derives from homage to two great warriors. The folk hero Guān Yǔ (关羽, 163–219), a great warrior in the Three Kingdom Period (ca. 184–280), was deified in the sixth or seventh century and has many temples in his name in the Tài Mountain region. Similarly, Zhāng Sānfēng (张三丰), a great Daoist priest and the creator of the Chinese martial art Tài Jí Quán, is associated with Wǔdāng Mountain, on which he lived. In traditional Daoist culture, Guān Yǔ is considered a protective spirit and Zhāng Sānfēng is an immortal. However, the religious significance of these attributes has been lost, and most modern Chinese would not even recognize the Daoist style of the buildings (de Bruyn, 2004).

As this evidence indicates, many Daoist texts, buildings, and sacred sites remained after the Cultural Revolution, but most common Chinese people no longer understood their religious meaning (Little & Eichman, 2000). As such, they are memory structures that have lost their usefulness as aids to Chinese collective remembering. Even more germane to our argument is that many Chinese still perform Daoist rituals but are unaware of their Daoist significance. Although "knowing how" to perform a ritual could be construed as a form of procedural (implicit) collective memory, rituals also have declarative (explicit) aspects. Most Americans, for instance, "know that" we carve jack-o-lanterns, dress up in scary costumes, and go trick-or-treating on Halloween. In parallel with the mainland Chinese vis-à-vis Daoist rituals, most Americans do not "know that" Halloween is the eve of All Saints' Day, a day designated by the Roman Catholic Church to honor saints and martyrs. But this parallel between Daoism and Catholicism is imperfect because, unlike the mainland Chinese, some Americans are aware of the religious connection and advocate for the religious

2. These observations are based on author Zhang's 6 years of life in Kūnmíng.

perspective (e.g., Ankerberg et al., 2008). Those who advocated for an overtly religious perspective during the Cultural Revolution in China were silenced. We will reflect more on this loss of continuity in the next section. Here we note that folk Daoism was still widely practiced in China after the Cultural Revolution, but its religious meaning, a declarative memory item which otherwise would have been transmitted and consolidated by religious specialists acting as relaters, was lost.

Daoism in Modern Chinese Literature in Mainland China

The removal of the interpretive community of priests for Daoist rituals and practices from common Chinese life does not mean that the people forgot Daoism altogether. What was lost were mainly the authoritative interpretations for Daoist rituals and practices that imbued them with religions significance, the "why" that went with the "how." Other important aspects of Daoism had become so integrated into Chinese thought that they are still alive in the Chinese collective memory of Daoism. The works of Ah Chéng (阿城, 1949–), a well-known representative of the Thinking Generation, illustrate the persistence of memory of Daoism. For Ah Chéng, the cultural lives of people in the remote countryside continue to exhibit an unpolished purity and faithfulness to Daoism.

Ah Chéng gained instant recognition among Chinese readers in 1984 with the publication of *The Chess Master* (棋王), which the official Writers' Association of China named the "best novel of the year." The subsequent publication of *The Lord of Trees* (树王) and *King of Children* (孩子王) raised his reputation still higher. He was asked to give talks at prestigious literary conferences and invited to visit Hong Kong and the United States. Numerous Chinese magazines and newspapers introduced him to their readers, and many well-known writers and scholars of Chinese literature have enthusiastically recommended his works to the public.

Like Ah Chéng himself, the narrator of the *The Chess Master* is a high-school graduate sent to a state farm in a southwestern forest during the Cultural Revolution. On the train he meets another city youth, Wáng Yīshēng (王一生). The story then centers around Wáng and his love for chess. Ah Chéng describes Wáng's experience living on the state farm in a poetic discourse with a romantic voice. He conjures a nostalgic world that recalls the leisure often found in traditional Chinese essay and poetry, but in the novel this leisure is a consequence of the balanced, Daoist way of life as lived by the mountain people.

Wáng illustrates the equilibrium of the material and the spiritual in order to come to terms with man's "original nature." This nature consists of two contrasting but complementary components: "biological needs" (生, e.g., the need for food) and "spiritual needs" (性, the obsession with something not directly tied to biological needs, like chess). The relationship between these two components instantiates the complicated interaction between antithetical binaries in Daoist cosmology: in opposition but not completely antagonistic, one depending upon the other, clashing as well as cooperating in order to realize their fullest potential as a unity (Ying, 1992). Thus, the "biological needs" and "spiritual needs" form a Tàijí (太极), the basic principle of the Daoist cosmos that Westerners recognize in the yin–yang symbol. With its focus on balance and harmony, Ah Chéng's novel, and its enthusiastic reception by readers, illustrate how the 1000-year-old Daoist views of the world still persisted in the Thinking Generation's memory in the 1980s despite the Cultural Revolution.

Another novel, *The Lord of Trees* is a eulogy for "the man of nature," Xiāo Gēda (肖疙瘩), who fights the destruction of a forest on a state farm. From the perspective of an educated city youth, Lǐ Lì (李立), Xiāo represents what is generally thought to be the traditional "Eastern mind," which abides by the law of the Great Universe, the "Great Chain of Being" idea (cf. Ellen, 1986). Xiāo embodies the Daoist ideal that humans should follow the flow of nature rather than try to counteract it. When urban youths are sent to cut down all the "useless trees," Xiāo tells them, "This tree cannot be cut down . . . Even the lord of trees will not cut it . . . It has become a spirit." For Xiao, humans should not remove trees simply because they think trees are useless for agriculture. Rather, everything in the world exists for a reason, whether or not it accords with people's agendas. To defend the trees from human invasion is to defend a way of life, a tradition, a philosophy, and a religion almost exterminated by the Cultural Revolution. But to the urban youths, this philosophy is stupid and backward. They finally cut down the tree of the lord. Xiāo then falls sick and eventually dies.

The story of Xiāo is an example of the persistence of Chinese collective memory of religion after the Cultural Revolution. It declares, in a lyrical and imaginative yet explicit way, the canonical nature of Daoism. The Daoist ideal is for human beings to embody the Dào such that it becomes the whole of the person. Disembodiment means death because it removes a person's source of life. By creating Xiāo as a man of nature who embodies Daoist ideals for humans, "A [sic] Chéng conveys his longing for the Daoist ideal of life" (Ying, 1992, p. 88).

What was lost and what persisted of mainland Chinese collective memory of Daoism can also be revealed by comparison with the Daoist experience of the Chinese in Northern Thailand KMT villages. The KMT villagers practice more of the Daoist rituals than the mainland Chinese. Moreover, they maintain a fuller living memory of Daoism because it is enlivened by its continuity with the past. As anthropologist Chāng Wen-Chin has documented (W. Chāng, 1999), Daoist specialists there preserve their connections with the common people by officiating at rituals and providing authoritative interpretations. For example, spiritual leaders in the villages still make red cloth (Fortune Cloth) containing charms and Daoist talismanic symbols, and most families hang a piece of this cloth in their home to discourage evil spirits. In addition, Daoist priests lead the villagers' performance of the annual ritual of Cháodǒu (朝斗), in which they worship the six personified deities of the Big Dipper. For funerals, people invite Daoist celebrants to preside over salvational rites, in which they lead one of the three souls of the deceased to the underworld, direct one to stay at the grave, and invite one to inhabit the family altar. Moreover, during the burial, villagers make a sacrifice to the Dragon Veins of the mountain to protect the tomb and the soul of the deceased.

The KMT villagers also maintain a memory of the religious meaning of Daoist stories. For example, the story of "The Journey to the West" (西游记) tells how Monkey Sūn, under the instruction of a Daoist immortal, becomes a great warrior and escorts Xuán Zàng (玄奘), the most well-known Buddhist monk in the Táng Dynasty, to India to seek the Buddhist Sutras. This story is well known in both Northern Thailand KMT villages and in mainland China. In Thailand, the story is not only told but performed in village exorcisms to drive off misfortunes, as the great warrior Sūn wards away evil (W. Chāng, 1999). On the mainland, although the story has been filmed and has become a children's favorite, it is only a story; people look for fun in it but have forgotten that the story has a deep religious significance and was also a model for exorcisms in imperial China since the time of the Míng Dynasty.

"The Journey to the West" provides a clear example of how mainland Chinese have retained a remote collective memory item, that of "the story of Monkey Sūn," but have lost the more recent collective memory that ties the story to its traditional past. It is precisely the continuity with the past, or of the relationship of recent events (births, marriages, deaths, rituals, and performances) with already consolidated memories (such as a "Daoism" schema), that was disrupted as a result of the removal from society of its social hippocampus. The Cultural Revolution could not erase the already consolidated items from collective memory (core philosophy, favorite stories), but because the Communists caused such trauma to organized Daoism, the relationality element (the priests in the case of Daoism) could not establish the connections necessary for the next generation of mainland Chinese to understand the meaning of the consolidated forms that persisted. Our model thus accounts for the differences in the collective memory for Daoism between mainland and refugee Nationalist Chinese.

The Trauma's Effect on Buddhism

Buddhism was introduced into China from India in the first century CE. Becoming an integral part of traditional Chinese culture, it reached its zenith in the Suí and Táng Dynasties (581–907). Buddhist monks are credited with introducing prosaic items such as the chair and sugar to China, as well as philosophical concepts such as transcending the reality of suffering (Kieschnick, 2003). Buddhism began to decline in the late Sòng Dynasty (960–1279), both because of the moral degeneration of the monasteries (e.g., monks purchasing certificates and honorary master titles from the government) and because of attacks against Buddhism by increasingly influential neo-Confucianists (K. Chen, 1964). The Communist movements in the twentieth century actively accelerated this decline. Part of the Communists' antipathy toward Buddhism was aroused by Buddhist doctrines advocating nonviolence and equality among humans and other creatures, which represented a strong psychological impediment to the Communists' dream of changing the world through violent class struggle (Welch, 1968).

To control Buddhists politically and spiritually, the CCP established the Chinese Buddhist Association (CBA) in 1953. The ostensible goal of the CBA was to unite all followers of Buddhism under the leadership of the CCP; the real goal was to eliminate Buddhism in China. The CBA began by "reeducating" Buddhists according to Maoism and by drawing them into the Communist movement. To transform Buddhists into secular Chinese citizens, the CCP further decreed that they had to vote, hold office, and serve in the army. The real clamp-down came during the Cultural Revolution: All religious associations, journals, and bookstores were shut down. All religious activity of Buddhist priests, as well as the "spontaneous" practices of the masses, were forbidden. By the end of 1976, all Buddhist monasteries in China's metropolitan areas were closed, and statues of Chairman Máo replaced destroyed Buddha figures. Buddhism had reached its nadir in the history of China; as sinologist Holmes Welch observed, "[F]rom the point of view of the outside observer, Buddhism had disappeared as completely as if it had been swallowed up in a black hole of anti-matter" (Welch, 1968, p. 344).

Against the near-total destruction of their material culture and faith practices, Buddhists struggled to maintain some semblance of a spiritual life through a syncretism of Buddhism and Marxism. Arguing that Buddhist doctrine was neither superstitious nor reactionary but scientific and compatible with Marxism, Buddhists tried to negotiate a more lenient policy toward the religion. For example, Zhào Pǔchū (赵朴初, 1907–2000), who later became the president of the CBA, said in 1957: "The goal of Buddhism was to benefit living creatures and beautify the land," and "since the communist party had the same goal, Buddhists were cooperating with it." According to him, Chairman Máo exemplified the Buddha mind for his dream of "serving the people," and the fact that the army functioned without insignia of rank was a sign that the Communist Party had attained Buddha's ideal of equality (Welch, 1968, pp. 347–357).

Religious leaders also attempted to propitiate the CCP with offers to exploit Buddhism for diplomatic purposes. During 1952–1966, the CCP did in fact overtly appropriate Buddhism in its interactions with Southeast Asian Buddhist countries. Its aim was to show that China was not an alien country but shared a common religious tradition with its neighbors. To take a recent example, the PRC uses the long (official) association between Chinese and Tibetan Buddhism as part of the justification for its claims on that vast mountain territory (Welch, 1968). Buddhism thus continues to provide a convenient pretext and a comfortable (if somewhat disingenuous) atmosphere for conciliating Asian leaders and other visitors.

Despite the hopes of religious leaders, this propitiation did not save Buddhism in mainland China. Buddhism as a living religion disappeared at the end of the Cultural Revolution. The large number of Buddhist monasteries and temples that once dotted the landscape were converted to other uses. Although a small number of culturally important temples remained opened as museums and parks, and old monks were employed there as custodians; no monks were allowed to perform Buddhist rites for the laity or to lead religious practices. Devotional clubs, Buddhist journals, and bookstores were not revived—except for one or two that were needed to show to visitors. As with Daoism, Buddhist activities privately carried out at home lost the guidance from monks and nuns that was necessary to maintain their deeper meaning and connection with the larger community of Buddhists. Even after the Cultural Revolution, the CCP continued to discourage popular Buddhist worship, and modern, Western education continued to reduce the number of worshippers. Though some Buddhist art and literature were preserved in museums and libraries, for common people they became mere artifacts.

The Persistence of Memory of Buddhism in Modern Chinese Literature

Although Buddhism as an organized religion disappeared from mainland China, Welch assured his readers 30 years ago that "Buddhist metaphors continued to be part of the language and Buddhist ideas like karma and rebirth will not be expunged from the popular mind" (Welch, 1968, pp. 379–380). This section will illustrate the stable memory of Buddhism by analyzing how Wáng Anyì, a well-known female writer in the 1980s and 1990s and now a professor at Fùdàn University (复旦大学) at Shànghǎi, has used the Buddhist notion of foreordination in her writing. She borrowed the plot structure of a well-known love story in the Táng Dynasty to write about a modern woman's love and life in China from the 1940s to the 1980s. The popularity of *The Song of Everlasting Sorrow* (长恨歌, 1995), one of her best-loved novels, demonstrates the tenacity of Buddhist thought in Chinese collective memory.

Wáng Anyì was born in 1954 in Nánjīng and raised in Shànghǎi, where she is currently chairwoman of its Writers' Association. After a brief period in rural China as an educated urban youth during the Cultural Revolution, she returned to Shànghǎi to work as a magazine editor. Wáng first came to prominence as a root-seeker with works such as *Love in a Small Town* (小城之恋, 1986), *Love on a Barren Mountain* (荒山之恋, 1986), and *Love in Brocade Valley* (锦绣谷之恋, 1987). She has achieved international recognition with her novels, novellas, and collections of short stories, which have been translated into English, German, Dutch, French, and Czech. In 1990, her novella *Little Bao Village* (1985) won the 10th *Times* Book Award. In 1998, *The Song of Everlasting Sorrow* was awarded the prestigious (Chinese) Máo Dùn award (茅盾文学奖). It was recently translated into English (A. Wáng, 2008).

The Song of Everlasting Sorrow has the same title as a widely known poem by Bái Jūyì (白居易, 772–846). This poem made a lasting contribution to the discourse on qíng (情, eroticism, love, passion) that developed in the literary circles of the mid-Táng Dynasty (618–907). The traditional verse (on which Wáng's novel is based) tells the true story of the relationship between Táng Xuánzōng (唐玄宗, Emperor Xuánzōng of the Táng Dynasty, 685–762), and Yáng Guìfēi (杨贵妃, Imperial Consort Yáng, 719–756). Their love exceeded Confucian bounds, distracted the emperor from governing and thereby indirectly resulted in the Rebellion of An Lùshān (755–763). When An (安禄山, 703–757) raised an army and began capturing cities, the court was forced to flee to southern China. Ministers and soldiers blamed Yáng for causing the rebellion and requested that the emperor execute her, threatening that they would not follow him otherwise. He finally agreed, allowing a palace eunuch to take her to a Buddhist shrine and strangle her.

Even after the court successfully put down the rebellion, Yáng's death became the source of "everlasting sorrow" for the emperor. Xuánzōng abdicated the throne in favor of his son and lived the remainder of his life in grief. The real love story ended there, but Bái Jūyì extended the story in the poem he wrote about it. He imagined that a Daoist, having traveled to an island of the immortals, took back to the emperor a gold hairpin and a shell box from Yáng. These two tokens, while not reuniting the lovers, reminded the emperor of their vow of unfailing love. The poem ends with these lines: "Ancient Heaven and Old Earth in time will cease / only this sorrow goes on without end" (天长地久有时尽,此恨绵绵无绝期).

This love story illustrates the Chinese notion of foreordination, which was introduced to the country through Indian Buddhism and then indigenized. When the Chinese were slow to accept Buddhism, early Buddhist missionaries (re)interpreted the Buddhist notion of foreordination in terms of a basic principle from the *Yì Jīng*, with which the Chinese were already familiar. The *Yì Jīng* is an ancient Chinese divination book and the most self-contained description of the Chinese notion of the workings of the universe and human societies. The relevant principle here is that when a thing reaches its extremity, it will become its opposite (物极必反). In this case, when the state of love reaches its peak of happiness, it will result in disaster and convert to its opposite, regret and suffering. In the Buddhist interpretation, the ultimate fate of such intense, fanatical love is inescapable because it is foreordained.

Wáng Anyì's novel *The Song of Everlasting Sorrow*, in casting the traditional poem in a modern context, illustrates that modern Chinese have retained the concept of foreordination. Wáng adopts the theme of eros (*qíng*), as exemplified in Bái Jūyì's poem, and uses the love between Yáng and the emperor to plot the story of Wáng Qíyáo's love for a person of high position and her tragic death as its consequence. The fictional Wáng Qíyáo's only asset is her beauty, and it procures for her third place in the first Miss Shànghǎi beauty pageant after the end of the Second Sino-Japanese War (1937-1945). This parallels the story of beautiful Yáng's election as imperial consort. The competition brings Wáng Qíyáo to prominence in the public eye and to the notice of Lǐ Zhǔrèn, a high military official of the KMT. Just as Yáng became an imperial consort, Wáng Qíyáo became Lǐ Zhǔrèn's mistress, and Lǐ Zhǔrèn monopolizes the sexuality of Wáng Qíyáo, which, as "emperor," would be his prerogative.

The parallel here is not perfect (see also Xiao, 2004). The love between Yáng and emperor Xuánzōng leads to a national crisis and Yáng's execution, whereas the civil war between the KMT and the CCP is contemporaneous but independent from the love between Wáng Qíyáo and Lǐ Zhǔrèn. Moreover, in contrast to Yáng's story, Lǐ dies (in a plane crash) while Wáng still lives, but these discrepancies do not mean that Wáng can escape her destiny. After the death of Lǐ, Wáng has to make her way through the vastly altered world of Communist China without the protection Lǐ once provided. All her assets—her beauty and position as the third Miss Shànghǎi—were denounced in the Communist period as capitalistic remnants. Nor does this hard life save Wáng from Yáng's fate. In the poem, a palace eunuch strangles Yáng at a Buddhist shrine; in the novel, Wáng is strangled in her bed by an acquaintance who wants to steal the few remaining pieces of the jewelry Lǐ had given her.

The interval between the two stories is more than 1000 years, and the respective cultural and social situations are completely different, but the notion of foreordination enables Wáng Anyì to connect these two separate stories and to highlight the tragic consequences of passionate love. Tragedy is predetermined by the nature of love in Chinese culture: it must culminate in despair. As do Ah Chéng's works discussed earlier, Wáng Anyì's novel exemplifies the kind of evidence that supports our hypothesis of collective retrograde amnesia. It declares, poignantly and imaginatively yet explicitly, an ancient aspect of Buddhist philosophy. Although the continuity that would have connected the idea of foreordination with its religious origin as part of Buddhism had been disrupted, the idea itself persisted and resurfaced as the driving force for an influential piece of literature. The centuries-long presence of Buddhist concepts in the Chinese collective memory buffer provided ample time for the social hippocampus to relate them with other items of Chinese thought and consolidate them as a component of the broader Chinese social framework. What was lost were the memories of the events that link these living concepts with their proper Buddhist past. The continuous existence of the idea of the nonpermanent state of love in Chinese society demonstrates the stability of the religious notion of foreordination, which survived the Cultural Revolution.

Memory for Buddhism in Táiwān and Northern Thailand KMT Villages

In Táiwān, Buddhism competes for adherents with Confucianism, Christianity, and indigenous religions. Nevertheless, the persistent efforts of Buddhists to exert a beneficial influence on people's lives serves to promote a living memory of Buddhism among Táiwān Chinese. Institutionalized Buddhism in Táiwān was introduced from the mainland in the early seventeenth century when General Zhèng Chénggōng (郑成功, 1624–1662) led Chinese from Fújiàn and Guǎngdōng (广东) to expel the Dutch from Táiwān (Jiāng, 1996). These Chinese brought a particular brand of Chinese Buddhism to Táiwān, known in Chinese literature as Jiāng-Zhè Buddhism, which emphasizes organized chanting and incense-burning in temples by monks, nuns, and laypeople. Táiwān Chinese combined Daoism and Buddhism with indigenous spiritual practices to develop a Táiwānese folk Buddhism known as the vegetarian religion (斋教), in which adherents cultivate themselves at home without abandoning the secular life.

In the early twentieth century, the Japanese initiated a transformation of Táiwānese Buddhism. In 1895, the Qīng court ceded Táiwān to Japan as tribute after the First Sino-Japanese War (1894-1895). The Japanese vice regal government considered traditional Chinese religious beliefs, centered around temple practices, as the main barrier to their policy of "turning the Chinese population of Táiwān into ethnic Japanese citizens" (C.B. Jones, 1999). They then launched a "temple restructuring" movement to raze temples, shrines, and meeting halls in order to acculturate Táiwān Chinese to Japanese beliefs and values. Chinese Buddhist temples had to submit to the Japanese Buddhist establishment. As a result, when Japan returned Táiwān to China in 1945, Táiwān Chinese viewed Buddhist temples as appendages of Japanese Buddhism.

Government-supported Buddhism in Táiwān began in 1949, when the KMT government retreated to the island. Monks from the Nationalist government's transplanted Buddhist Association of the Republic of China (BAROC) quickly took over all Buddhist establishments in Táiwān and occupied all previously confiscated Japanese temples. Over the next two decades, these monks utilized government power to establish a new systematic monastic ordination and completely monopolized Buddhism in Táiwān (C.B. Jones, 1999; Jiāng, 2003). Despite these efforts, the BAROC failed to solve the problems caused by rapid economic development since the 1980s. The rising civil society called on nongovernmental Buddhist organizations to create a modern "pure living land." The Pure Land refers in the first instance to an imaginary cosmic realm of abstract Buddha Consciousness, but it can be used as a model for actual life here and now (Lín, 2003). Through this nongovernmental route, Buddhism found its way into modern Táiwānese society.

Learning from the spread of Christianity, Buddhists started to establish organizations in schools, to preach throughout the island, to publish Buddhist journals and pamphlets, and to appear on television. To take some specific examples, Master Xīng Yún (星云) blends Buddhist text-exegesis with tourism, educating people through recreation. Master Zhèng Yán (证严) established the Cí Jì (慈济) Foundation, in order to realize the Buddhist dream of serving people. And Master Xiǎo Yún (晓云) founded a college to teach modern science and technology together with ancient Buddhist doctrine (S. Lǐ, 2006). These efforts have met with great success, and Buddhist leaders have set up an extensive Buddhist center in Mázhú Yuán (麻竹园) to serve the increasing numbers of people in Táiwān who are adopting the faith. In the 1990s, Buddhism in Táiwān became one of the most well-developed forms of Buddhism in the world, and Buddhist thought has penetrated many aspects of modern society (Jiāng, 1996).

In comparison with Buddhism in Táiwān, Buddhism in Northern Thailand KMT villages has had a smoother ride. Buddhist specialists in Northern Thailand continue to lead villagers in rituals such as the salvational funerary ritual that guides the soul to the afterworld, and villagers strongly believe in samsāra—that death is followed by another life. The worship of Guānyīn (观音), the Goddess of Mercy and one of the most popular Buddhist deities in Chinese Buddhism, is widespread in Northern Thailand. Every KMT village has a Guānyīn temple where people hold an annual festival to celebrate Guānyīn's birthday and her attainment of perfect life (W. Chāng, 1999). Thus in Northern Thailand and, despite its rockier history, in Táiwān as well, Buddhism now seems to exist as an integrated whole in the sense that there is no separation between existing Buddhist concepts and their Buddhist origins. As for Confucianism and Daoism and so also for Buddhism, continuity with the past in Táiwān and Northern Thailand but not on mainland China indicates relative loss of recent religious memory in China, even though central religious concepts live on in all three places. These findings support our hypothesis that destruction of the religious social hippocampus on mainland China during the Cultural Revolution resulted in a pattern of collective retrograde amnesia that is analogous to that of individual retrograde amnesia: loss of recent but retention of remote collective memory.

Conclusion

To conclude this chapter, we argue that after the Cultural Revolution, the three major religions did disappear on mainland China, but that basic religious notions—such as the Confucian ideals of human nature, the Daoist view of human beings and their integrated position in the universe, and the Buddhist view of foreordination—remained in an accessible and useable form in mainland Chinese collective memory. We suspect that other central religious themes have also survived as already consolidated, stable collective memories, and further anthropologic and historical analysis of the mainland Chinese will uncover them. The basic religious notions that we identified as surviving the Cultural Revolution belong to three different religious traditions, but they share a common quality that could be described as gist-like—as capturing a central theme while allowing the specifics to escape. Such generalization accompanied by loss of attribution is characteristic of consolidated, stable memory (see chapter 7).

Generalization of religious concepts probably occurs in all societies (as all societies have some form of religion; see Boyer, 2001). For example, many Americans would associate The Golden Rule with Christianity but could not cite the chapters and verses from the Bible in which The Golden Rule is expounded. In contrast, American Christian specialists including clergy could easily do so. These specialists also "remind" the populace of the religious origins and significance of their well-established (i.e., consolidated) cultural values and practices. Religious specialists on mainland China, who composed the religious social hippocampus, were heavily suppressed during the Cultural Revolution, and the main result was a break in continuity between stable (and still current) concepts and their religious origins. In either case, with or without recent trauma, remote religious collective memory exists in generalized form, as we would expect from our model in which stable memory formation involves generalization.

Where Americans (and Táiwān Chinese and Northern Thailand KMT villagers) differ religiously from mainland Chinese is that the former know the religious significance of their surviving notions (e.g., foreordination for Buddhism or The Golden Rule for Christianity). The comparison between the two Chinese diasporic populations—Táiwān Chinese and Northern Thailand KMT villagers—and mainland Chinese shows that collective memory for some of the central tenets of religions, established in the remote past, have survived the Cultural Revolution. What has been destroyed in China is memory for the recent events that would have lighted the path that connects their current experience to their traditional past. Our analysis of religion provides strong evidence in favor of our hypothesis that, by removing the mainland Chinese social hippocampus, collective memory consolidation was disrupted, resulting in a collective retrograde amnesia in which remote memory was spared while recent memory was lost. This pattern of retrograde amnesia is analogous to that observed after destruction of the actual, neurologic hippocampus in individuals (see chapter 2).

Religions have ancient origins. For that reason, they were appropriate subjects of study in this chapter in which we focused on the persistence of already consolidated, stable memory. Our goal in the next chapter is to demonstrate that the Cultural Revolution, as a social trauma directed specifically at the relaters in Chinese society, prevented the consolidation of events that occurred, and were still labile, during the pretrauma period (from 1911 to 1949; see chapter 9). In chapter 11 we will, therefore, focus on developments in literature, including the writing of poetry but also of novels and short stories, activities that were in full swing during the pretrauma period in mainland China. Collective memory for literary works, authors, and events was indeed diminished by the Cultural Revolution, strongly suggesting that this trauma caused a failure of consolidation for recent memory items.

11 Loss of Unconsolidated Collective Memory

In the previous chapter, we showed that mainland Chinese collective memory for well-established, already consolidated religious themes survived intact despite the trauma of the Cultural Revolution but that connections with the religious past did not survive, strongly suggesting that the process of consolidating these more recent events had been interrupted. This chapter presents further evidence for interrupted collective memory consolidation. It will center around the loss of memory of the Thinking Generation for three of the most prominent writers of the Peking school (Hú Shì, Zhōu Zuòrén, and Shěn Cóngwén), whom the Chinese Communist Party (CCP) traumatically targeted as antirevolutionary for their efforts to preserve traditional Chinese artistic influences in modern Chinese literature. Because a collective can regenerate its parts in ways that an individual cannot, the loss of collective memory for events of the pre-trauma period (1911–1949) is now being compensated. In the more liberal political atmosphere that the CCP has allowed in contemporary China, and with the help of non-mainland diasporic populations, a new "social hippocampus" has arisen on the Chinese mainland that has resumed the process of collective memory formation.

This chapter also examines in detail an example of disrupted (and later resumed) literary memory consolidation, focusing on works of fiction by author Zhāng Àilíng (张爱玲, or Eileen Chang, 1920–1995). This example is emblematic of the loss of recent collective memory due to destruction of an existing social hippocampus and its recovery under circumstances in which an equivalent collective relater appears to take its place. Zhāng produced most of her writings between 1942 and 1945 and achieved great fame in Shànghǎi during the pretrauma period, shortly before the CCP's takeover of the country. In the 1950s, the CCP denounced Zhāng's work as both feudal and capitalistic, and her books disappeared as the result of suppression during the trauma period. Zhāng's work was still labile when the Cultural Revolution began, and the CCP's destruction of the mainland Chinese social hippocampus prevented its consolidation as part of the stable collective memory of the mainland Chinese. The lack of awareness among mainland Chinese for Zhāng's fiction during the posttrauma period constitutes a case of collective retrograde amnesia for recent events. Liberalization in China since the 1980s has allowed the emergence of a new social hippocampus, and mainland Chinese have been reintroduced to Zhāng Ailíng's work. It has surpassed its former popularity.

The Thinking Generation's Memory for Hú Shì

Hú Shì (胡适, 1891–1962) was one of the founders of modern Chinese literature. He was admired as a cultural hero and revolutionary by the youth of the 1920s. He received a classical Chinese education from his mother as a child and then attended middle and high school in Shànghǎi, which at the time was the home of Chinese publishing and of a number of progressive, reform-minded Chinese journals. Like many of his peers, Hú received his university education in America in the heyday of the Progressive Era, when Pragmatism, new historiography, new jurisdiction, and literary realism were at their peaks. At Cornell and Columbia Universities in the 1910s, Hú came under the influence of American philosopher and education reformer John Dewey (1859–1952) and his philosophy of Instrumentalism (M. Li, 1990).

Instrumentalism proposes that concepts and theories are merely useful instruments whose worth is measured by how effectively they explain and predict phenomena. Applying Instrumentalism to the situation in China, Hú published "Some Modest Suggestions for Literary Reform" (文学改良刍议) in *New Youth* in 1917, advocating for a modern Chinese literature written in vernacular style and for the adoption of the vernacular as the national language. For Hú, writing in the vernacular provided an escape from the burdens of classical Chinese, whose Confucian elements he found to be barriers to modernization (M. Li, 1990). His hope was that vernacular writing would stimulate the reformation of traditional China.

New Youth, of which Chén Dúxiù was the chief editor, was the leading journal for Chinese intellectuals in the 1910s and 1920s. Chén embraced Hú's proposal and followed up with an editorial entitled "On Literary Revolution" (文学革命论) in the next issue of *New Youth*. These two articles launched a revolution in 1917, and that year is considered the beginning of modern Chinese literature. This linguistic and literary revolution spread widely throughout China via the May Fourth Movement and had tremendous social impact. Beginning in 1919, various vernacular magazines and newspapers were published and circulated. In 1920, the Ministry of Education ordered that classical Chinese be replaced with the vernacular in all grades of primary school, and later in middle and high school. By 1921, the vernacular had largely achieved both official and popular recognition as the national language (G. Zhōu, 2002). To emphasize just how important and influential Hú and Chén were, consider the following anecdote. When Máo Zédōng was asked sometime in the 1920s "What is your favorite magazine?" he answered, *New Youth*; and "Who do you admire most?" "Chén Dúxiù and Hú Shì" (Z. Wáng, 1989, p. 1).

To legitimize further the vernacular literary reform, Hú applied the methods of new historiography that he learned in the United States to the history of Chinese literature. In his *A History of Vernacular Literature* (1928), he proposed the revolutionary idea that vernacular literature should be considered as the core of Chinese literature, not as inferior to classical Chinese literature, and not as a degenerate linguistic variety used by lower-class people. Hú argued that the history of Chinese literature is a history of the slow transformation of classical Chinese and of the development of the vernacular. The May Fourth Movement actually sped up the organic and slowly progressing vernacularization of Chinese literature (G. Zhōu, 2002).

Though a radical himself in the 1910s and 1920s, Hú began to resist the coarse and dogmatic revolutionary literature thriving in Shànghǎi and what he saw as its irresponsible critiques of Confucianism (Huáng, 2002). For Hú, such critiques came from the authors' ignorance of the value of Confucianism in modern society. Despite his re-evaluation of classical Chinese language and literature, from the late 1920s Hú began to apply systematic methods to identify the benefits of Chinese tradition, hoping to construct a new Chinese culture that would be based on modern science while absorbing the merits of Confucianism. In Běijīng he founded the journal *Independent Commentary* (独立评论) to rally writers of a similarly independent mind, namely, writers who neither blindly accepted Western theories nor ardently clung to traditional Chinese culture. Gradually his own attitude toward Confucianism became increasingly positive, and this was encouraged by the new-found pro-Chinese sentiment that arose in the face of Japanese aggression in the 1930s (M. Li, 1990).

Although Hú's insistence on independent thinking made him favor neither Communism nor Nationalism, his educational background and social status led him to develop close personal friendships with many high officials in the Guómíndǎng (Chinese Nationalist Party, or KMT). In 1937, when the Second Sino-Japanese War broke out, Hú visited the United States as a government-sponsored scholar, an intellectual envoy sent to secure American support for China against Japan. His efforts were rewarded with conditional U.S. support of the KMT government in 1942. In recognition of his contribution to the success of the anti-Japan war, Hú was appointed as the chancellor of Peking University in 1946, from which post he tried to promote a free China. He condemned the civil war between the CCP and the KMT, but not, apparently, out of love for the CCP. In 1949 when the CCP took over mainland China, he left China for the United States and became the curator of the Gest Oriental Library at Princeton University where he continued his study of Chinese history and literature. In 1956, he moved to Táiwān and died there in 1962.

Unfortunately, Hú's acceptance of Confucianism and close relationship with the KMT were used as justification by the CCP in targeting him as a dangerous "capitalist element." In 1954, the CCP established a committee investigating and critiquing Hú's thought, which resulted in the publication of eight volumes of *Critiques of Hú Shì* in 1955 (Grieder, 1970). Hú became a negative figure on the Chinese mainland as a result of the Cultural Revolution. His role as a founder of modern Chinese literature was deliberately erased by the CCP.

Literary scholar Chan Ngon Fung (N.F. Chan, 2004) has described how Hú became a forgotten figure. One of the most popular textbooks of modern Chinese literature for college-level students, published in 1979 and reprinted 15 times until 1995, did not even mention Hú Shì in a total of 837 pages. Even worse, in *A History of Chinese Modern Literature*, the first college text on modern Chinese literature written in the spirit of greater openness after the Cultural Revolution, the authors claimed that the initiator of the New Literature Revolution was (only) Chén Dúxiù. Hú's first paper launching the literary revolution and his subsequent efforts on its behalf were omitted (Tián & Sūn, 1979). Clearly, collective memory for Hú Shì on the Chinese mainland had been lost as a result of suppression during the Cultural Revolution.

In contrast, Hú gained great fame in both the United States and Táiwān. In 1955, Hú was nominated by *Look* magazine (in the United States) as one of the world's most important people (Dé. Táng, 1991). In Táiwān, Hú was praised by the KMT government as a fighter against mainland Communism, despite Hú's occasional criticism of the KMT government's anti-democratic policies. Upon his death in 1962, commemorative events and essays reinforced the Táiwānese public's memory of Hú. The Táiwān KMT government declared Hú's philosophy as "steadfast and complete anti-communist thought" and underwrote his funeral. More than 20,000 people participated in the farewell. Jiǎng Jièshí, the president of Táiwān at that time, sent his deputy along with an obituary couplet (a eulogy in the form of a two-line poem). Hú's residence at the Táiwān Central Academy has been commemorated as a museum (Zhū, 1988). Today, Hú is still ranked by Táiwānese scholars as one of the most important Chinese cultural figures to appear in 3000 years of Chinese history (Dé. Táng, 1991). Unsurprisingly, Hú's death was not even officially reported on the mainland in the 1960s (Grieder, 1970).

Not until the late 1980s, when Cultural Fever flourished, did mainland Chinese begin to reevaluate Hú and recover their collective memory for him. In 1999, Chinese linguist Jì Xiànlín (1911–2009) published *Recollection and Reflection in front of Hú Shì's Tomb*. In the essay, Jì acknowledges Hú's "mistakes" as well as his contributions to modern Chinese literature. Jì had been persecuted during the Cultural Revolution but was subsequently rehabilitated by the CCP and by the late 1990s was ranked as one of the top scholars in the humanities in mainland China. Such bold statements from so respected a scholar helped to turn the tide of antipathy away from non-leftist authors. In the context of our model of memory consolidation, Jì Xiànlín was part of a regenerating mainland Chinese social hippocampus that reestablished relationships between the work of non-leftist authors of the pretrauma period and more traditional Chinese culture.

This section shows that Hú Shì and his work were very well represented in mainland Chinese collective memory during the pretrauma period, were virtually absent from mainland Chinese collective memory during the posttrauma period, and are gradually being reintroduced in contemporary China. In contrast, the previous chapter showed that religious principles were well represented in mainland Chinese collective memory during both the pretrauma and posttrauma periods, though the communities in Chinese society that served as the memory makers (i.e., relaters) for both literature and religion were severely suppressed by the CCP during the trauma period (which includes the Cultural Revolution). Our model offers an explanation for these findings. Hú Shì and his work were relatively recent and did not appear on the scene until after the pretrauma period began. During that period they were very well represented, but in a labile state, in the mainland Chinese collective memory buffer. The relaters for literary memory were actively processing Hú and his work and relating them to other authors and works past and present, but this process was disrupted before Hú and his contributions could be consolidated as stable collective memory items. In contrast, religious principles, which had been introduced centuries before, had already consolidated as stable memory items, so disruption of the relaters for religion did not remove religious principles from mainland Chinese collective memory.

The difference between the two cases is critically important, because the CCP wanted to erase both non-Communist literature and religion (and other components of traditional culture) from Chinese collective memory. To do this, they emptied the collective buffer and destroyed the collective relater, but this did not erase collective memories for religious principles because they had already consolidated. The CCP enjoyed greater success in using directed social trauma to erase mainland Chinese collective memories for recent literary developments and authors, because those had yet to consolidate. What we observed in this section for Hú Shì occurred for other authors as well, as we describe in subsequent sections.

The Thinking Generation's Memory for Zhōu Zuòrén

Chinese essayist, translator, poet, and cultural critic Zhōu Zuòrén (周作人, 1885–1967) was one of the central figures in the Vernacular Literature Reform Movement of the late 1910s. Zhōu proposed a Literature of Humanity (人的文学), dividing literature into "human/good" and "not-human/ bad." According to Zhōu, good literature expresses experiences or emotions that all humans share, regardless of nationality, race, or class, whereas bad literature is written under the influence of a religious or political ideology that does not promote the values of humanity. To capture humanity in literature, Zhōu developed the genre of the personal essay (小品文), writing about one's understanding and experience of the world in which "one finds significance to sustain oneself" (Gunn, 1980, p. 156). Such significance goes beyond the materiality and locality of personal experience to reach a level common to all human beings (Daruvala, 2000). In this sense, Zhōu's personal essays advanced the May Fourth agenda of building a "literature of the ordinary people." Consequently, Zhōu was considered a member of the Chinese literati and an iconoclastic cultural reformer along with other major figures associated with the literary journal *New Youth* such as Hú Shì, Chén Dúxiù, and his older brother Lǔ Xùn (X. Zhāng, 1995).

Communists did not hold a monopoly on literary repression in twentieth-century China. In the 1910s and into the 1920s, Běijīng (Peking) was the cultural center of China. In the mid-1920s, however, warlords Duàn Qíruì (段祺瑞, 1865–1936) and later Zhāng Zuòlín (张作霖, 1875–1928) established a repressive military government at Běijīng. In 1926, to control intellectuals and purge opponents, the military government banned several important newspapers and journals, such as the *Peking Gazette* (京报), *Modern Commentary* (现代评论), and *Yǔsī* (雨丝). It also listed 50 intellectuals whom they would arrest, including those with opposing views such as Lǔ Xùn, who supported the Communist view of literature as a tool for political struggle, and Zhōu Zuòrén, who strove to establish a literary world that was free of political influence. Though it should already be obvious to any student of history, this fact is one among many proving that the Communist Chinese were not the only powerful group ever to target a social hippocampus for the purpose of disrupting (or altering) its associated process of collective memory consolidation (see chapter 12). This specific example of suppression by Běijīng warlords forced many liberal professors and writers who advocated reform to flee to Shànghǎi. Soon, Běijīng lost its cultural importance and Shànghǎi became a new cultural center, particularly for revolutionary literature.

Although he was targeted by the military government, Zhōu chose to stay in Běijīng and distanced himself from the revolution brewing in southern China. This estrangement from the New Culture Movement was consistent with an ethic that also led him to produce a literature capturing individual understandings, tastes, and experiences of the world, rather than preaching doctrines concerning the development of society. In 1928, Zhōu published his famous "On Reading Behind Closed Doors" (闭门读书论), in which he renounced his popularly appointed role as a social, intellectual, and cultural critic. In the following decades, Zhōu lived in Běijīng as a social hermit and independent writer and refused to establish any partisan political alliances.

When class-struggle in China escalated into a prolonged military confrontation between the CCP and the KMT, and the entire cultural sphere in China came to be dominated by an intense political/ideological antagonism between "left" and "right," Zhōu founded the journal *Luòtuó Cǎo* (骆驼草, *Camel Grass*), where he published many beautiful and highly personal essays that promoted his alternative idea of modern literature (Daruvala, 2000). Then, after Japan occupied the capital city, Zhōu accepted a job from the puppet government. Drawing on his classical education, Zhōu took pleasure in translating Greek, Roman, Russian, and Japanese works into the Chinese vernacular. Unsurprisingly, Zhōu's works were viewed by radical left-wing writers as both "feudal" and "bourgeois" and as irredeemably "right" and "reactionary."

Zhōu's distance from revolutionary literature, his Japanese wife, and his collaboration with the occupying Japanese government sealed his fate in the future Communist China. In December 1946, shortly after the end of the Second Sino-Japanese War, the Supreme Court of the Republic of China sentenced Zhōu to 10 years in prison for his collaboration. He was freed by the CCP government in the 1950s, but the damage had already been done. In 1942, Máo Zédōng had chosen Zhōu's work as the official representative of bad literature in "Talks at the Yán'ān Forum on Art and Literature." According to Máo, good literature should motivate soldiers to fight against the Japanese and to prepare the masses for the inevitable civil war between the CCP and the KMT. To make the situation worse for Zhōu, Máo further denounced him as a writer of traitorous literature (X. Zhāng, 1995). With this epithet, Máo froze Zhōu's work out of consideration by literary relaters who might still have been in a position to integrate it before the severe crackdown of the Cultural Revolution. The ensuing suppression of the social hippocampus ensured that Zhōu and his work remained unconsolidated. The result was the erasure of Zhōu's reputation from mainland Chinese collective memory (Baird, 2006).

Evidence that collective memory for Zhōu's work failed to consolidate can be found in memory structures, such as textbooks, from the trauma period (1950–1976). For example, Zhōu's name was systematically removed from all textbooks on modern Chinese literature and was seldom mentioned on academic or public occasions. A Yīng (阿英 [first name "A", second name Yīng is a pen name for Xìng Qián]), the author of the widely read *A History of Late Qing Fiction*, had all references to Zhōu in the pre-1937 edition removed from the 1955 edition. And in *An Outline of Chinese New Literature* (1956), which was used as a textbook of modern Chinese literature in many universities from the late 1950s to the early 1980s, Zhōu's name did not appear at all, even in the chapters on the development of the Chinese essay, in which Zhōu played a vital role. In rare cases where Zhōu was mentioned, he was denounced as a negative figure in the history of modern Chinese literature. For instance, in *A Draft History of Chinese New Literature* (1951), which was the textbook of modern Chinese literature used in Peking University, one of the top two universities on the mainland, Zhōu was described not as a brave fighter against feudalism but as complicit in all the things that made China "backwards" before the CCP came to power (Zhāng & Zhāng, 1986, pp. 406–407).

Unlike Hú Shì, Zhōu's name was not known outside mainland China during the Communist movements. Therefore, we cannot compare memory for Zhōu on the Chinese mainland, where the CCP suppressed collective memory consolidation, with memory for Zhōu in areas not affected by the Cultural Revolution. However, a comparison of Zhōu's great fame in the first half of the twentieth century (including the pretrauma period) with his literary oblivion from mainland China during and immediately after the Cultural Revolution (the trauma and posttrauma periods) demonstrates how the CCP interrupted integration of Zhōu into the collective memory of modern Chinese literature. The suppression of Zhōu during the Cultural Revolution was so effective that it took mainland Chinese almost 10 years to recover their interest in him after the CCP's literary doctrines were abandoned.

Between 1979 and 1989, when the pendulum swung away from repression and in the direction of permissiveness, a "New Era" of inquisitiveness into Chinese culture and literature commenced. During this era, scholars intended to retrace and reinvent a Chinese "liberal" intellectual discourse in the systematic reading or rereading of the non-leftist intellectuals and writers. Slowly, many of Zhōu's previously prohibited works were reprinted. Between 1980 and 1990, more than 130 articles on Zhōu were published, and several biographies and critical works were produced (X. Zhāng, 1995). But when the mainland Chinese began to re-read Zhōu, "he often [had] to be introduced as Lǔxùn's [Lǔ Xùn's] brother . . . To many Chinese before the mid-1980s, Zhōu's literary production and intellectual activities during the 1930s [were] virtually unknown" (X. Zhāng, 1995). This lack of knowledge clearly demonstrates collective retrograde amnesia for a figure that otherwise would surely have been consolidated into the pantheon (schema or framework) of great Chinese writers of the twentieth century.

Qián Lǐqún (钱理群), a well-known and respected authority on the writings of Lǔ Xùn and Zhōu Zuòrén, even offered a course introducing Zhōu Zuòrén to graduate students and visiting scholars at Běijīng University in 1987. This was the first attempt on the mainland to systematically reintroduce Zhōu to the reading public. Students welcomed this course enthusiastically, and Qián had to change classrooms several times in order to accommodate increasing numbers of inquisitive and receptive students. Root-seeking writers similarly rediscovered Zhōu in the late 1980s and his writings inspired them to move beyond the formulaic literature of the Cultural Revolution (Qián, 1991). The intense interest in Zhōu and his writings in contemporary China strongly suggest that they would have consolidated into mainland Chinese literary memory in his time had the consolidation process not been disrupted. Although the Communists effectively suppressed his influence from the 1950s to the 1970s, from the late 1980s on Zhōu was successfully re-incorporated into collective memory, once relaters such as Qián Lǐqún were allowed to draw the public's attention to him again.

The example of Zhōu Zuòrén illustrates how items of collective memory held in a labile state in the collective memory buffer can fail to consolidate if the normal functioning of the collective relater is disrupted. Zhōu, his works, and their influence on the development of Chinese literature were very well known during the pretrauma period, indicating that they were very much present in the mainland Chinese collective memory buffer. Moreover, Zhōu was a controversial figure, indicating that he was also being actively processed as a potential memory item by existing collective relaters, but a stable memory for Zhōu failed to consolidate because that process of relationality was disrupted by the Cultural Revolution. That the consolidation process for Zhōu has resumed in contemporary China is a thread we retake in chapter 12.

According to our consolidation model, in which the buffer and relater are the first and second stages, respectively, it should be possible to prevent new memory items from consolidating by destroying the buffer rather than the relater. Clearly, new memory items cannot consolidate if they are prevented from entering the consolidation process in the first place, but it is likely more difficult to destroy the buffer than the relater. In individuals, the relater corresponds to the hippocampus, which is a relatively circumscribed region of the forebrain, whereas the buffer appears to involve the forebrain more broadly (see chapters 5 and 6). Analogously, the social hippocampus is restricted to a relatively small group of influential meaning-makers, whereas the collective buffer involves the society more broadly. Targeting the social hippocampus is an effective way to prevent recent, labile memory items from consolidating as part of stable collective memory. The CCP put this strategy to effective use in preventing well-known and influential but non-Communist writers from being represented in stable mainland Chinese collective memory.

The Thinking Generation's Memory for Shěn Cóngwén

If we were to rank modern Chinese writers according to their prominence during the pretrauma period, Shěn Cóngwén (沈从文, 1902–1988) would perhaps be second only to Lǔ Xùn (darling of the CCP). Shěn was born in western Húnán, a remote rural area in China's hinterland inhabited by the Hàn, the Miáo (苗), and other ethnic peoples. His paternal grandmother was a Miáo, and his mother was a Tǔjiā (土家; the Miáo and the Tǔjiā are 2 of 55 officially identified ethnic minorities in China). Between the ages of 14 and 20 years, Shěn worked variously as a soldier, clerk, tax collector, and staff secretary, as well as at a printing company, an orphanage, and the recruiting station of an army warlord. This background exposed Shěn to the complexity and diversity of real lives, and his lived experience nourished his work.

Shěn was a passionate experimentalist in fiction writing. In short stories such as "Bǎizǐ" (柏子, 1928), "The Love of a Shaman" (神巫之爱, 1929), and "The Long River" (长河, 1938), he used both foreign modernist and traditional Chinese modes of literary expression (M. Chéng, 1999). This made Shěn a figure of controversy for the CCP. Before 1949, Shěn was censured by left-wing writers for his middle-of-the-road political position and liberal attitude toward literature. Some of his works were criticized for being unrealistic and politically naive (Y. Wáng, 1951/1986; Zhāng & Zhāng, 1986). After 1949, Shěn was forbidden to write, which led to a failed suicide attempt. He regained emotional equilibrium but he never returned to fiction writing. He later became an authority on the costumes of the Miáo and ancient Chinese.

In the People's Republic of China (PRC), Shěn, like Hú and Zhōu, was *literati non gratis*. Consequently, several influential textbooks on the history of modern Chinese literature published in the 1950s and 1960s either did not mention his name and work or classified him as a "bourgeois writer" and distorted his views (M. Chéng, 1999). In *A History of Modern Chinese Literature* (1979), for instance, Tián and Sūn spent only 3 of 544 pages on Shěn. Though praising him by noting his exceptionally broad social understanding, they complained that Shěn's writing was rambling and vacuous: "If we condense his one thousand words into several hundreds, we still get his meaning" (Tián & Sūn, 1979, p. 254). By contrast, the authors devoted two long chapters each to Lǔ Xùn, whom the CCP praised as a literary hero, and Guō Mòruò, a proponent of Marxist literature.

So profound was the resulting retrograde amnesia for Shěn that even his once-acclaimed love story, "The Frontier City" (边城, 1933–1934), exerted almost no influence on the Thinking Generation's love stories, where we do find traces of the influences of other May Fourth writers (Louie, 1985). Consequently, when Shěn was rediscovered with other non-leftist authors in the late 1980s, he was known not as a writer but as a scholar of Miáo costumes and Chinese antiquities. Not until more interest developed in Shěn's research on antiques as vessels of traditional culture did the Chinese begin to rediscover Shěn's literary merit as a brilliant and sensitive interpreter of the human condition (Líng, 1991; Wah, 1988; M. Chéng, 1999).

In contrast to the oblivion they suffered on the mainland from 1949 to the late 1980s, Shěn's works were internationally recognized and praised as great modern Chinese literary creations. For example, according to *A*

Bibliography of Studies and Translations of Modern Chinese Literature 1918–1942 (1975), there was an average of one dissertation or book on Shěn completed or published every two years in the United States and Australia. That is an amazing level of New World output, considering the degree to which Shěn and his experimental fiction had been withheld from the collective memory of his own country.

We should emphasize here that the CCP's reasons for suppressing the three writers we have just discussed were twofold. On the one hand, these writers held literary views at odds with the CCP. Hú's belief in independent thinking went against the CCP's desire to control Chinese minds; Zhōu's literature of humanity directly contrasted with Máo's political literary doctrines; and Shěn's literary experimentation challenged the CCP's formulaic proletarian literature. On the other hand, these writers embraced traditional cultures in China, which the CCP had deemed "backward." From the late 1920s Hú praised Confucianism; Zhōu's personal essays exhibited a Daoist view of life and the world; and many of Shěn's works were devoted to the Shamanism of the Miáo in Húnán Province.

Although these writers were very well known during the first half of the twentieth century, the CCP prevented the Chinese reading public from integrating them into their memory of modern literature. They arrived on the scene too late relative to the onset of the Cultural Revolution—the crackdown occurred before their influence had consolidated. This led to reduced mainland collective memory for the three writers after the Cultural Revolution. The robust collective memory of these writers off the mainland (except Zhōu Zuòrén) confirm an inference from our model of memory consolidation, that if the CCP had not suppressed this trio, then the Chinese probably would not have forgotten them or their works. The data on the mainland Chinese people's enthusiastically revived interest in these writers after the late 1980s confirm this inference.

Aborted Memory Consolidation: Zhāng Ailíng and Her Work (1942 to mid-1980s)

Before the May Fourth and vernacular literary movements, young writers pushing the boundaries of Chinese literature were members of the classical Mandarin Duck and Butterfly literary schools, both based in Shànghǎi.[1]

1. The mandarin duck and the butterfly were frequently used as symbols for unhappy lovers in traditional Chinese literature: Lovers forbidden to live together would become mandarin ducks or butterflies so as to live together after their deaths.

When the Japanese occupied the city in December 1941, many writers fled to Chóngqìng (重庆) or Kūnmíng in southwestern China, but some Butterfly writers stayed in Shànghǎi. Hoping to offend neither the Japanese rulers nor their Chinese compatriots, they wrote stories removed from the current political milieu: nostalgic essays, historical anecdotes, and vignettes of the small but delightful events of daily life or the dazzling lifestyle of a Chinese metropolis (Y. Chén, 1998). In this politically complicated but artistically fertile context, Zhāng Ailíng began to publish.

Zhāng Ailíng's fiction first appeared in 1942. She published mostly in Mandarin Duck and Butterfly journals, including *Violet* (紫罗兰) and *Phenomenon* (万象), but her work was written in the vernacular language and style. Her two stories "Aloeswood Ashes (沉香屑): The First Incense Brazier" and "Aloeswood Ashes: The Second Incense Brazier," published in *Violet*, gained wide popularity in Shànghǎi literary circles. Zhōu Shòujuān (周瘦鹃), the editor of *Violet*, commented that these stories were reminiscent of *The Dream of the Red Chamber*, one of the four greatest traditional novels representing the highest achievement of Chinese fiction (Y. Chén, 1998). More generally, her work merged the May Fourth writers' exploration of human nature and moral/spiritual confusion in an unstable society with traditional folk literature, more common to the Butterfly school, describing fallen, urban aristocratic women and tacky townsfolk facing the great social transformation of China in the early twentieth century (S. Chén, 1999). Because of her seamless fusion of the timeless and the quotidian, Zhāng's name became the emblem of modern Chinese folk literature, expressing classical themes in a vernacular spirit, and her work was welcomed by both traditional and May Fourth writers.

Zhāng deliberately wrote about humankind rather than political ideology. In her own words: "I even just write about trivial things between man and woman. My works contain neither war, nor revolution" (A. Zhāng, 1944/2001, p. 74; cf. Gunn, 1980; D. Wáng, 2001). Accordingly, after the CCP took over Shànghǎi in 1949, Zhāng was pushed aside in favor of more propagandistic authors. In 1952, she left Shànghǎi for Hong Kong, and then in 1955 she moved to the United States. Beginning in 1966, she taught at the University of California at Berkeley, introducing Chinese literature to the English-speaking world.

Despite her popularity and critical acclaim during the pretrauma period, Zhāng and her work were almost completely forgotten on the mainland until the late 1980s. Yú Qīng and Jīn Hóngdá, two Táiwānese experts on the scholarship of Zhāng's work, note that "from 1949 to the 1980s, the study of Zhāng Ailíng on the mainland has been blank" (Yú, 1991, p. 1).

More specifically, Chinese literary scholar Chén Zǐshàn (陈子善) found that even in the 1990s, though mainland scholars had begun to understand the preceding three decades of scholarship on Zhāng from outside the PRC, they had only a very vague idea of how Zhāng's contemporaries received her work in the 1940s. Such ignorance illustrates that mainland Chinese after the Cultural Revolution not only forgot Zhāng as an important writer, but they also knew little about the influences she exerted on her contemporaries, who would have been the ones to consolidate her as an item of collective memory. In other words, Zhāng and her contemporaries *were* the literary social hippocampus. Because they were suppressed during the trauma period, the memories they would have consolidated failed to consolidate.

The failure to consolidate Zhāng's work due to suppression of the social hippocampus is brought out clearly by literary scholar Kē Líng (柯灵). Kē has observed that Zhāng's work was not mentioned in almost any of the history books of modern Chinese literature before the late 1980s because "there was no place for her in the history of modern Chinese literature: The May Fourth literature in the 1920s aimed to fight against imperialism and feudalism, the revolutionary literature in the 1930s aimed to reflect the class struggle, and during the Sino-Japan war, literature was about anti-Japan" (Kē, 1984/1991, p. 10). From the standpoint of our memory consolidation model, it is telling that Kē explains the lack of memory for Zhāng as a lack of established relationships between her work and that of others during the pretrauma period. But we have already seen that, during the pretrauma period, Zhāng was clearly connected to the May Fourth Movement and the long tradition of folk literature through her use of the vernacular style and her interest in the lives that common Chinese people actually lived. Zhāng's work could have been placed into a system of literary relationships by her contemporaries, but they, as the social hippocampus for literature in their time, experienced the suppression of the Communist movements.

In contrast to scholars on the mainland, scholars of Chinese literature off the mainland ranked Zhāng as one of the two most important writers in modern Chinese literature (the other was Lǔ Xùn) (Yú, 1987, 1991). From the 1950s, when Zhāng's story "The Rice-Sprout Song" was first introduced in Táiwān, professional writers' interests in and ordinary audience's increasing curiosity about Zhāng developed into the "Zhāng Ailíng Phenomenon," and by the 1970s and 1980s, Zhāng's works were celebrated in Táiwān (Hoyan, 1996; Z. Yáng, 1999; Y. Chén, 1998; F. Chén, 1999). A succession of authors learned from and imitated her. In *A History of Modern*

Chinese Fiction (1961), pioneer Táiwānese Zhāng scholar Tsi-an Hsia spends 42 pages on Zhāng, compared with only 26 pages for Lǔ Xùn, praising her as "the most excellent writer from the May Fourth Movement" (Hsia, 1999, p. 289).

Although many other scholars believe that Hsia exaggerated Zhāng's contribution, there is no doubt that Zhāng takes a prominent place in Táiwān Chinese collective memory, and her name has been written into the history of Táiwānese literature (F. Chén, 1999). In 1994, Zhāng was awarded the 17th annual Special Achievement Award from the *China Times* newspaper in Táiwān. In 1995, when the news of her death was reported in Táiwān and Hong Kong, commemorative essays appeared in newspaper columns and literary magazines for nearly half a year. By 2005, there had been more than 10 master's theses on Zhāng in Táiwān. And in university courses concerning modern Chinese literature, Zhāng has been continuously ranked in the same position as Lǔ Xùn (Z. Chén, 2003).

On the mainland, the CCP aborted the process of Zhāng's integration into the Chinese people's memory of modern literature. It took the Chinese reading public almost 10 years to resume memory consolidation for Zhāng after the Cultural Revolution. Unsurprisingly, when her works were reintroduced on the mainland after the late 1980s from Táiwān and Hong Kong, "many Chinese considered Zhāng as a writer from Táiwān or Hong Kong, rather than a mainland writer" (Yú, 1991, p. 1). Because they had never heard of her before, they could not associate Zhāng with her own native country! Yet the "Zhāng Ailíng Phenomenon" did spread to the mainland after the mid-1980s. Many of her important works have been made available, and they have influenced professional mainland writers (Hoyan, 1996). Now, when a mainland writer such as Xū Lán (须籣), who has been called the "little Zhāng Ailíng," shows an ability similar to Zhāng's in creating a splendid but desolate social landscape, we can be assured that Zhāng has become a part of mainland Chinese culture, albeit half a century later than she would have been if mainland collective memory consolidation had not been disrupted.

Conclusion

The data in this chapter confirm that labile, non-Communist pretrauma memory items associated with the writers Hú Shì, Zhōu Zuòrén, Shěn Cóngwén, and Zhāng Ailíng were absent from mainland Chinese stable memory after the Cultural Revolution. Importantly, the authors we examined here were not only famous in their time, but they were also founders of modern Chinese literature. The rebound in interest in these writers in contemporary China, aided by improved communication with non-mainland Chinese populations, is testament to their importance and strongly suggests that these labile memory items would have become stable items if the process of collective memory consolidation on the mainland had not been disrupted. Our analysis suggests that the cause of this disruption was the Cultural Revolution. This Communist movement affected the entire population, but it disproportionately affected the "elites," which we identify as the mainland Chinese social hippocampus, the meaning-makers of the society. Our analysis suggests that the CCP deliberately targeted them in order to change the course of Chinese collective memory consolidation.

In this chapter, we examined mainland Chinese collective memory for literary figures that arrived on the scene during the pretrauma period. We then searched in memory structures (such as textbooks) and in researchers' reports for evidence of their existence in mainland posttrauma collective memory and found little, suggesting that a retrograde amnesia for recent events had in fact occurred. In contrast, we showed in chapter 10 that posttrauma memory for remote items (i.e., religious principles) is spared. Taken together, these findings verify our prediction that a collective retrograde amnesia, characterized by loss of recent but retention of remote memory (see figure 2.6), should result from trauma to the relational element of a society. Verification of this prediction provides support, on collective levels, for our three-in-one model (see figure 4.2), from which the collective retrograde amnesia prediction was derived.

Our three-in-one model proposes that memory consolidation involves buffering of labile memory items, interrelating of selected labile and some stable items, and generalizing a stable representation from those interrelationships, all under the influence of the consolidating entity (see figure 4.2). As we discussed in chapter 7, the generalization step is not absolute: some items enter stable memory intact, without losing their unique characteristics. The result is a stable memory construct that contains specific memory items properly positioned within a generalized framework. It was this sort of stable memory construct that we considered on a collective level here in part III, in the contexts of Chinese memory for religion and literature.

In chapter 10, we found evidence for retention of collective memory for general religious principles in the posttrauma period. According to our model, the fact that these religious principles were represented in a general way implies that they were part of stable memory, because generalized memory occurs only as stable memory in the model, but it does not imply that stable religious memory did not also contain discrete memory items. On the contrary, Confucius almost certainly existed on the posttrauma Chinese mainland as a distinct item of stable collective memory, although the connection between Confucius and the Confucian religious principles appropriated by the CCP was hidden from the people.

In this chapter (chapter 11), we found evidence of loss of collective memory for creative writers as distinct memory items. This is also consistent with our model because retrograde amnesia occurs only for recently acquired items, which occur only as labile (and so disruptable) memory items in the model. But this does not imply that memory for literature would only ever be stored as discrete items. On the contrary, as these authors and their work re-enter the mainland Chinese collective memory buffer, and as collective relaters in contemporary mainland China establish interrelationships between them and other aspects of Chinese culture, they will form a stable memory construct in which the names of Hú Shì, Zhōu Zuòrén, Shěn Cóngwén, and Zhāng Ailíng will be associated with a generalized conception of a writing style that connected real Chinese people with timeless Chinese values.

Of course, not everything that enters the buffer is selected, related, and consolidated. The goals of the consolidating entity are critical, and these changed over the long period we considered here: In the first half of the twentieth century, traditional Chinese culture was treated as either a tool for, or a barrier to, China's modernization, whereas the cultural rebound in the late 1980s aimed to (re)establish an "authentic" Chinese identity. The different conditions surrounding the demise and the resuscitation of traditional culture suggest that the recovered memories probably differ significantly from those that would have formed from uninterrupted collective consolidation. Nevertheless, the chapters of part III demonstrate that during the interval between the end of the Cultural Revolution and the late 1980s, when the Thinking Generation started root-seeking, there was reduced memory for recent but not for remote people and events. This finding is consistent with our prediction that collective retrograde amnesia should follow destruction of the social hippocampus and should affect mainly recent memory, just as the individual retrograde amnesia does that follows destruction of the neurologic hippocampus (see figure 2.6).

An alternative interpretation of the collective amnesia findings we present in this chapter, which would not support our model, is that people forget authors and their writings anyway, whether or not social trauma occurs. This is tantamount to stating, for example, that Americans care more about who wins the Superbowl than about who wins the National Book Award. But there are two reasons to reject this alternative interpretation. First, any collective is composed of many groups, each with its own social hippocampus, and we would expect, in any large society, that different memories would be consolidated by different groups. The evidence we present in this chapter demonstrates failure of consolidation by the literary community itself, which is the proper relater for literary memory, due to its suppression on mainland China by the Communist movements of the CCP. Second, the robust rebound in interest in the writers we cover in this chapter clearly proves that their work is memorable and is now actively being consolidated into the collective memory of the mainland Chinese. These efforts at recovery provide further, strong evidence that the trauma of the Cultural Revolution did indeed disrupt the process of collective memory consolidation.

12 Conclusions

The previous chapters have provided support for our argument that (1) the process of memory consolidation—by which we mean the formation of stable, generalized knowledge constructs—is the same on individual and collective levels; (2) observations made on the individual level apply on collective levels and vice versa; and (3) retrograde amnesia as observed on the individual level can also be observed on collective levels. Our proposed consolidation model has four elements that work together to form memory representations. The first and third elements, respectively, handle the labile (short-term) and stable (long-term) memories that constitute, respectively, the raw material and end products of the consolidation process. To this well-established pair of components we add an intervening relationality element (selector/relater) and an encompassing entity; the latter functions in the model as a meta-element. We have provided evidence from the existing literature for these four consolidation components on both individual and collective levels. We have also argued for the recursion of stable memory and for the influence of the remembering entity on the consolidation process. Finally, we have presented evidence that disruption of collective memory consolidation, due to traumatic injury of the social relater, causes a collective retrograde amnesia that affects recent but spares remote collective memory. In this concluding chapter, we indicate ways in which our model can contribute to the study of memory on different levels. Specifically, we suggest that considering the constructive influence of the remembering entity would broaden scientific understandings of individual memory, and that considering the formation of memory through consolidation would deepen humanistic understandings of collective memory. Our memory consolidation model provides a new conceptual framework that can organize findings, facilitate reasoning, stimulate new insights, and propose testable hypotheses about individual and collective memory.

This concluding chapter begins by discussing (re)construction and (re)consolidation from the viewpoint of individual memory sciences. Because of the constructive and recursive idea of consolidation we hold, we must reject reconsolidation as it is usually defined. Then we switch from individuals to collectives and address the perceived conflict between history and collective memory. We suggest instead that history may be viewed as a form of collective memory produced specifically by historians, who as a collective entity have goals that differ in certain respects from those of the larger society. This chapter continues with an interpretation of James Wertsch's *Voices of Collective Remembering* (2002) in the context of our model that uses the concepts we have been developing throughout this book: a consolidation process that makes stable representations from labile input, the recursion of generalized knowledge, and the influence of the remembering entity. The chapter closes with a discussion of the loss of memory on collective levels and the potential to recover it, thanks to interaction between collectives and the regenerative capability of successive generations.

Construction in Individual Memory

We have described three basic elements of memory consolidation—the buffer, relater, and generalizer—and a fourth, the entity, which is a meta-element that encompasses the other three. An important property of the model is the influence on the consolidation process both of the recursion of generalized knowledge and of the needs, goals, and emotions of the entity. Thus, consolidation is not a unidirectional progression from buffer to relater to generalizer; rather, generalized memories that have already been created can recur to influence the consolidation of new memory items. In addition, the core consolidation process can be biased by the entity's needs and goals, modulated by its emotions, and molded by its desire for a coherent narrative that is usable because it makes consistent sense to the entity.

Our three-in-one model is an extension of a two-element model proposed by McClelland and coworkers (1995), who combined a Hopfield network and a multilayer perceptron that act as a buffer and generalizer, respectively (see chapter 2). The buffer receives new items and accepts already learned items from the generalizer, and the buffer presents new and already leaned items, in interleaved fashion, to the generalizer for (re)learning. This combination allows the generalizer to learn new patterns (items) without losing (overwriting) the already learned patterns, thereby avoiding catastrophic interference (see also chapter 2). Together, these systems are more than the sum of their parts; the two systems interact harmoniously and overcome each other's weaknesses. The buffer can learn verbatim patterns in one exposure but has a limited capacity, and the generalizer can generalize over a potentially huge number of patterns but is slow to learn and prone to interference. The buffer and generalizer work together continually to incorporate the new without forgetting the old, and the result is generalized memory that has up-to-date information as well as the benefit of experience.

We proposed the three-in-one model as an extension of the model of McClelland and colleagues to explain a broader range of memory phenomena. Prominent among them is the observation that memories can depart considerably from reality. Such departures are initially due to the entity, which can introduce inaccuracies into the memories that are consolidated, and these inaccuracies can be compounded through recursion of consolidated memory back onto the ongoing consolidation process. Each core element of the three-in-one model is active during the consolidation process: the buffer holds new inputs and items recovered from generalized memory; relationality confers associative meanings; and generalization classifies items and codifies their relationships. The entity can influence all of the core elements. In conjunction with the generalizer, it can tag new items through attention or add old items to the buffer, but the entity can also rearrange items to suit its various purposes. Thus, memories produced through the consolidation process are nothing like verbatim records. They are indeed constructed.

Entity effects and generalized knowledge recursion influence remembering as well as memory formation. William Brewer and others describe remembering as building a "mental model" guided by "schema" (Brewer, 1987; Chinn & Brewer, 1996; Mishra & Brewer, 2003). Other literatures address entity and schema effects on memory formation and remembering in the context of false memories (e.g., Hyman & Loftus, 1998). Through convincing manipulations, such as retouched photos or the enacted and meretricious recollections of close relatives, experimenters create false memories by leading a human test subject to "remember" an event that never actually occurred. All of these authors are the intellectual successors of Frederic Bartlett (1932), whose generative schema theory proposed that memories are not veridical copies of events but constructed versions of events that are reassembled at the time of remembering, under the top-down influence of prior knowledge and the control of the remembering entity.

To situate our model within the memory literature, we must continually distinguish between memory, remembering, and memory formation. In simplest terms, "memory" corresponds to the contents of the buffer (labile memory) and generalizer (stable memory), plus whatever storage capacity the relater may have, whereas remembering and memory formation are processes involving an interaction among these three basic consolidation elements (buffer, relater, and generalizer) and between them and the entity. An autobiographical narrative, which is a generalized memory construct, illustrates this interaction as it concerns remembering. According to our model, an individual relating her life story is unlikely to "read" it directly out of her generalizer. She would more likely download remembered items from the generalizer to the relater, perhaps also introduce new items from the buffer, and she (as an entity) would weave (re-relate) them into an account that will be heavily influenced by her circumstances, particularly her relationship with her listeners and the effect she wishes her story to have on them (cf. Olney, 1998).

This view of remembering is consistent with that of Bartlett (1932) and Bruner (1990), in which "remembering" is akin to "thinking"; it is an active, constructive process that references prior representations created by memory consolidation but is not limited to them. The narrative process of remembering can innovate and even confabulate using the same sorts of cognitive mechanisms that are involved in creating a fictional account. In remembering by narrative construction, the past is an "organized mass" full of gaps and distortions, and a cogent story is created at the time of remembering to meet present needs (Bartlett, 1932).

"Narrative" is both a knowledge construct that resides in the generalizer and a process of remembering that arises from an interaction between the three basic memory elements under the influence of the entity. Because the three core elements are involved in both remembering and memory formation, a newly remembered autobiographical narrative is grist for the consolidation process in the sense that it is yet another version of the individual's life story that is generalized-over in forming stable memory. Our model accounts for the widely accepted phenomenon that our stories change as we tell them (see also chapter 7). Scientific evidence from humans and other animals is consistent with this common understanding.

There is a wealth of literature on (re)construction in remembering, especially autobiographical remembering, and in understanding patient histories (Freud, 1900/1953; Bartlett, 1932; Neisser, 1967; Sacks, 1985/1998; Bruner, 1990; Schacter, 1995). The phenomenon of memory distortion provides much of the proof of this narrative-based reconstruction of memories. Studies such as the one undertaken by psychologist and legal expert Elizabeth Loftus (1997) set out to implant a false memory of a vivid but fictional childhood event (such as being lost in a mall or knocking over the punch bowl at a wedding). After a series of interviews, more than 25% of subjects reported "recovering" the false memory, and many added elaborate details (Gentner & Loftus, 1979; Loftus & Pickrell, 1995; Loftus, 1996b). These and other "recovered memories" of unlikely events, such as alien abduction or satanic ritual abuse (Lynn & McConkey, 1998), bolster the notion of remembering as a constructivist concept, where the creation of an autobiographical narrative need not reference actual events. In some subjects, the reconstruction can have greater authority than the representation that existed before the false memories were introduced.

Our model suggests that false memories can be formed because memory consolidation is a continuous process in which the relationships between old and new memory items are constantly being reestablished. The overarching idea that remembering affects memory is not new. In his 1988 article on self-knowledge, American psychologist Ulrich Neisser asserts that a memory is best understood not as a unitary representation but as the result of four facets: the actual event, the subjective experience of the event, the formation of a memory, and the recall of the event. Except for the actual event, all of these facets are brain-derived and therefore prone to subjectivity. Critically, the recalled event, which may be inaccurate, becomes a new substrate for consolidation. As Bartlett's (1932) repeated recall experiments demonstrate, additions, commissions, exaggerations, and the more prosaic assumptions made in the representation for consistency's sake during mental construction become codified in the generalizer. All of this decreases the accuracy of any subsequent acts of remembering, but, from the entity's viewpoint, veridical recall is not necessarily the sole purpose of memory.

The results of experimental studies of what has been termed "reconsolidation" have been interpreted to mean that stable memory is not stable after all (Nader et al., 2000; Nader, 2003). Some researchers suggest that when a stable memory is recalled, it again becomes labile and therefore disruptable. Reconsolidation research begins with the work of psychologist James Misanin and colleagues. They reported in 1968 that rats that had successfully consolidated a fear response (presumably rendering memory for the response invulnerable to disruption by electroconvulsive shock; ECS) were once again vulnerable to amnesic intervention after being briefly exposed to the original response-producing experimental setup. Specifically, thirsty rats were placed in a box and allowed to drink water from a spout. Then an audible tone was played, after which the rats received electric shock to their feet. Normally, rats learn a fear response from this one trial such that, when placed back in the box, their rates of water-spout licking decrease sharply after the tone is played again. This learning is disrupted if ECS is administered immediately after the trial in which the tone is followed by foot shock. What Misanin and co-workers found was that if rats were later placed back in the box and then ECS was administered, learning of the fear response was disrupted as much as if the ECS had been administered immediately after the original learning trial (Misanin et al., 1968). This suggested to them that the learned fear response, which presumably had consolidated into a stable form, was recalled when rats were placed back in the box, and that this recall made the learned response labile again and therefore vulnerable to disruption by ECS. Their results suggested that the original fear response learning had been lost.

More recently, neurobiologist Karim Nader and his colleagues (Wang et al., 2005) have re-created some of these findings with protein-synthesis blockers as the memory-disrupting agent. The larger implications of this work seem to be that stable memories are made labile again when they are "activated" by recall. In addition to radically shaking up the more traditional view of stable memory, these reconsolidation findings open up potential new areas for psychiatric intervention. For instance, psychiatrist Roger Pitman has developed a treatment for posttraumatic stress disorder (PTSD) that involves administration of the drug propranolol, a consolidation-disrupting agent, while the patient is remembering the traumatic event (Pitman et al., 2002).

Despite their continued popularity and apparent practical usefulness, reconsolidation theories have come under suspicion in recent years. These theories assert that consolidation essentially runs in reverse whenever a previously stable memory is recalled, but the relevant evidence does not conclusively support that assertion (Riccio et al., 2006). Bartlett's original idea of schemata, and our model, suggest an alternative explanation for the reconsolidation results. When an old memory is activated—as it is in the case of rats being re-exposed to a fear-inducing experimental setup or of humans recounting their traumatic memories—the previous event is *reconstructed*, not simply recalled. Disruption of the reconstruction process can make the remembered version dramatically different from nondisrupted recollections. The new, disrupted reconstruction does not entirely overwrite the old representation; rather, elements of the two are melded together by the ongoing consolidation process; this necessarily changes the representation for the overall memory of the event or situation. Future reconstructions of the event can differ dramatically from the one made before the disruption and may therefore make it appear as though the first version of the memory had been "lost."

Support for this viewpoint comes from studies of hypothermia-induced amnesia in mice published in the same year as the original reconsolidation experiments (Lewis et al., 1968). Although the mice did exhibit the reconsolidation phenomenon, in that previously stable memories seemingly could be rendered disruptable by re-exposure to the original setup, their memories were not *totally* disrupted as they would have been by amnesia-causing treatment during initial learning. Instead, the mice could re-learn the previous memories more quickly than they could learn a new memory because the previous representation had not, in fact, been deleted: It was still available to facilitate re-learning. In a related study (Richardson et al., 1982), a brief re-exposure to the original setup that included the treat (water) but excluded additional disruption (more ECS) could also restore the supposedly disrupted memory. These experiments suggest that nondisrupted recollections can "wash out" the effects of a disrupted recollection on the overall memory for a situation, as the recursive consolidation process continually forms and reforms its stable, but modifiable, representation.

The results on false memory and memory reconsolidation underscore the fact that "stable," when used in reference to a type of memory, is a relative term. It might be better to refer to "labile" and "stable" memory as "more changeable" and "less changeable." Our model captures this dual, enduring-yet-malleable aspect of long-term memory that is stable but constantly reshaped by the consolidation process, as both new material and newly recalled and reconstructed memories are integrated with existing representations. Thus, generalized memory—constructed by the buffering, relationality, and generalizing elements and influenced by the entity—recurs to affect the three basic elements' consolidation work. Consolidation is best understood as a process that continually reshapes "less changeable" memory in a constant, recursive loop of reconstruction (recall) and reconsolidation (reformation).

History as a Type of Collective Memory

Issues of stability and accuracy are as important to collective memory scholars as they are to individual memory researchers. This section considers the distinction that is often made between "collective memory" and "history" and why our model supports the view that history should instead be seen as a type of collective memory. This view contrasts with that of many historians who place "history" in opposition to "collective memory." Beginning with Maurice Halbwachs (Halbwachs & Coser, 1925/1992) and continuing up to the present with scholars including Pierre Nora (1984, 1989), Noa Gedi and Yigal Elam (1996), Peter Novick (1999), and James Wertsch (2009), "collective memory" is seen as culturally biased and false. Its foil, "history," is considered universal and true. Wertsch (2009, p. 127) characterizes "collective memory" as subjective, originating from a single committed perspective, reflecting a particular group's framework, and anti-historical. In contrast, he describes "history" as objective, distanced from any particular perspective, and reflecting no particular social framework. This view gives the impression that "history" is at war with "collective memory," or at least that the two are in serious conflict.

Any conflict between "history" and "collective memory" can be resolved by seeing history as an instance of collective memory. We recognize both labile and stable forms of all instances of collective (and individual) memory, but for the purposes of this discussion we can emphasize the stable forms that are the products of their respective collective (or individual) entity's memory consolidation process. Thus, if collective memory is the (stable but ever-changing) product of the consolidation process of a specific collective entity, then history is the product of a consolidating collective entity composed of historians. What separates history from other types of collective memory are the influences brought to bear on the consolidation process by a group of historians acting as an entity. Our view contrasts sharply with the one that considers history to be objective and reflective of no particular social framework or perspective. If history is formed by a collective entity composed of historians, then it will be influenced by the social frameworks that guide that entity, as well as its goals, needs, desires, and even emotions. And if history is a form of collective memory, then like any other form of memory, individual or collective, it is subject to some inaccuracy. Simply put, history is what it is because historians actively make it that way, and if history is more accurate and verifiable than other types of collective memory (e.g., group identity narratives), then that is because these attributes are important to historians through some combination of the social frameworks and non-mnemonic factors (goals, plans, desires, emotions, etc.) that they share as a group.

We view history as a (stable but ever-changing) form of collective memory, but we certainly do not claim that there is no difference between representations of the past that are consolidated by entities composed of historians on the one hand and of a society at large on the other hand. As we explained in chapter 7, we make a distinction between "analytic history" and "popular history," which are consolidated by historians or the general public, respectively. And we certainly do not mean to disparage history (or analytic history) by placing it in the category of collective memory. We view collective memory very broadly: Under the heading of collective memory we include cultural schemata, identity narratives, foundational myths, religious belief systems, scientific paradigms, and both analytic and popular histories. We consider all of these as types of (stable but ever-changing) collective memory because they are all formed by a process of consolidation, in which a set of labile items are held temporarily, relationships are established between them, and a stable representation is distilled from that system of relationships, all under the influence of the consolidating entity.

Analogously, under the heading "welded metal constructions" we could include a sculpture by Alexander Calder and the chassis of a new Ford Fiesta. Obviously, these objects are not the same, but the process of construction that produced them is essentially the same, as metal components had to be collected, had to be placed in a certain spatial relationship, and had to be joined together under intense heat. Moreover, they were constructed for different reasons by entities that had different goals, plans, desires, and emotions. Reversing the analogy, we would not say that an analytic history and a popular history are the same, but we would say that the process of consolidation that formed them is, on an abstract level, essentially the same. We believe that seeing analytic and popular history as two different forms of consolidated collective memory will facilitate scholars' treatment of these consequential knowledge constructs.

Before going any further, let us acknowledge what should be apparent, that most historians no longer seek to write universal history. It is by now a commonplace that in the liberal and radical social, political, and intellectual trends of the 1960s and 1970s, "history" lost its innocence. Multiculturalists, postmodernists, hegemony theorists, feminists, social constructionists, deconstructionists, postcolonial theorists, and others who fit no easy label questioned the objectivity of academic history, the power dynamics of the Ivory Tower, and where the boundaries of knowledge of self and other fall. Historians now draw from a variety of backgrounds to write histories from myriad points of view.

Considering the diversity in outlook among modern practitioners of history, historian David Blight (2009) laments that:

If hardly anyone believed anymore in the "noble dream" of neutral narratives, then the subject might as well be competing narratives – memories and counter-memories – and how they emerge in relation to social and political power. (p. 241)

But what, we wonder, is so bad about that? As we have seen in the consolidation of scientific and other types of collective memory, competition among individuals drives the establishment of the systems of relationships that are then generalized-over in forming (and reforming) stable memory (see chapters 4 and 7). Competition likewise occurs between groups within a larger collective as it strives to consolidate a shared version of its past. The importance of competition to the process of collective consolidation leads us to predict, by analogy, that the neurophysiologic equivalent of "contest and negotiation" also occurs in individual brains as the relater constructs the webs of associations that are essential for individual memory consolidation. Whether between neurons or networks in individual brains, or between individuals or groups in collectives, competition between different systems of relationships is a natural concomitant of the memory consolidation process. If analytic history is largely accurate, then it owes part of that accuracy to competition among historians as separate individuals or between the different camps, factions, schools, and persuasions that compose their profession. This competition serves to diminish the effects of bias and the influences of the social frameworks to which historians are subject as individuals embedded in societies.

Like scientists, historians have an identity separate from society but are simultaneously part of society. Just as scientists wish to make their findings (once they are sure about them) available for use by the larger society, so historians wish to inform and instruct the larger society concerning its past and its relationships with the pasts of other societies. Many historians see their role as aids to collective remembering and desire to participate, as members of their respective communities, in acts of communal remembering. Historian Jay Winter (2009) conceives of "historical remembrance" as an amalgam of history and collective emotions such as those associated with deep interpersonal bonds and sacredness. It is the combination of these various factors, he states, which makes historical remembrance "a phenomenon of enduring power" (p. 255), and he points out that historians participate with non-historians in collective remembering. Indeed, rather than standing indifferent to the larger society out of disdain for its popular history, it may be that the desire to promote and participate in "historical remembrance" is what got many historians into the business in the first place!

Soviet-Era History Textbooks as a Case Study of Collective Memory

The study of shared views of the past can be facilitated by considering history as a type of collective memory and by focusing on collective memory *consolidation* rather than on collective memory per se. We can use our model to demonstrate how some of the existing work on collective memory could be seen in the context of collective memory consolidation. For instance, James Wertsch's *Voices of Collective Remembering* (2002) has added immensely to the theorization of collective memory. Wertsch explores narratives as cultural tools in Soviet and post-Soviet school textbooks, and he describes how modern states control the formation of collective memories by controlling the narratives of history. His analysis comes up against the problem of separating "history" from "collective memory," but to us it provides a unique view on a specific instance of collective memory consolidation as it is influenced by a powerful entity.

The confusion over whether to call the products of official propaganda machines collective memory or history can be alleviated by focusing not on the products but on the producers. As in China under the Chinese Communist Party, the Soviet Union essentially installed a single "social hippocampus"—the organizations of the Soviet Communist Party—to write and disseminate a version of history that reflected the Soviet narrative and its perceived needs. This version no doubt falls short of what many historians would consider analytic history, but the Soviets attempted to make it part of the popular history of Soviet society. The extent to which they succeeded tells us something about the collective consolidation process and the influence of the collective entity.

Although constrained by Marxist-Leninist frameworks, public debates about history continued in the Soviet Union from the time of the October Revolution in 1917 until 1931 when, as part of his "revolution from above," Secretary-General of the Communist Party Josef Stalin (1879–1953) effectively gagged the leading anti-Marxist historian, Mikhail Pokrovskii (1868–1932). "Stalinization" had come to the discipline of history. Under the false banner of the "Pokrovskii school," the Communist Party launched attacks in the late 1920s against non-Marxist historians and ways of viewing the past. Stalin (and/or the party) simultaneously expressed concern about history education in the schools, and the mid-1930s were spent devising a national curriculum and standard history texts for primary and secondary schools (Artizov, 1993). The authors commissioned to write the texts missed many deadlines, as each new manuscript proved unsatisfactory to Stalin's changing historical tastes. He was in the process of monopolizing his power as the Supreme Leader, and key to his success was a new, nonrevolutionary narrative for Soviet society.

History textbooks almost always represent the political and historiographical status quo (yes, even American ones; see Ward, 2006). The new status quo Stalin wanted to consolidate was one of a united, socialist republic. Whereas Pokrovskii and others had lambasted the tsars and all the social inequality and oppression they stood for, by 1937 Stalin wanted to celebrate the tsars' ability to unify the peoples and territories of the Russian Empire-*cum*-USSR. He was clearly seeking legitimacy in history for the federated state structure he wished to impose in the present. To signal further this unity, the resulting textbook, A.V. Shestakov's *Short Course of the History of the USSR*, which Stalin himself edited (naturally), was printed in the 15 languages of the ethnic peoples Stalin had designated within the federation.

The very changeability of the content of Soviet school history texts epitomizes our argument about the malleability of meaning of even stable, consolidated collective memories. For instance, in the 1930s, Stalin ordered that individual heroes be added to the collectivist narratives then common to Communist ideology, because these would make history more accessible to budding young socialists (Vähä, 2002). In the terms of our model, he focused his influential attention to select certain individuals to embody Soviet ideals from among the vast throng of the Soviet people. However, each successive Soviet premier (i.e., Nikita Khrushchev, Leonid Brezhnev, and Mikhail Gorbachev) changed the party narrative and its ideals, requiring the heroes to be recast with every change in ideology. This phenomenon is well illustrated by the example of a particularly hard-working shoveler known as Aleksei Grigorievich Stakhanov (1906–1977).

The eponymous Stakhanovite Movement was a Soviet economic and propaganda campaign that began in 1935 when it was reported that Aleksei Grigorievich Stakhanov regularly exceeded his coal-mining quota by 5%—and on one shift exceeded it by 14%. With the right tools, training, and creative accounting, Stakhanovites in all industries sped up their production for love of country and party. Stakhanov himself became a stable memory item, but the meanings associated with him changed over time. Thus, historian F. M. Eliisa Vähä writes,

All of the images of Aleksei Stakhanov were shaped by the most important pedagogical targets of the curriculum in each era. During Stalin's era these were patriotism and physical strength; in Khrushchev's era—the superiority of the socialist state over the west; in Brezhnev's era—collectivism and the goal of educating an unselfish socialist who had a strong will to be part of production; finally, in Gorbachev's era—individualism and initiative. (Vähä, 2002, p. 550)

Like Abraham Lincoln in the United States, Stakhanov was a distinct, stable memory item, and he appeared in schoolbooks for five decades. But the meaning attached to his name was malleable; it is as if the authors of each new edition related his name (an old, stable item) to the current ideologically appropriate heroic identity (a new, labile item) and then consolidated (or re-consolidated) this new pair for the next generation of schoolchildren. The same recasting of the past can be observed in the meanings attached to such stable collective Soviet memories as the October Revolution: Over time, political revolution was increasingly confined to the period before the Second World War, whereas the post-war era was defined as one of increasing economic prosperity and technological advancement. In each new edition of the common textbook, the sections had to be slightly rewritten to reflect a gradually changing political narrative (Evans, 1976).

Exactly the same recasting of the contributions of "the great heroes of an earlier age" occurs with each new textbook of science, as described in Thomas S. Kuhn's *The Structure of Scientific Revolutions* (1962). Kuhn writes,

Partly by selection and partly by distortion, the scientists of earlier ages are implicitly represented as having worked upon the same set of fixed problems and in accordance with the same set of fixed canons that the most recent revolution in scientific theory and method has made seem scientific. (Kuhn, 1962/1996, p. 138)

Kuhn, of course, conceived of "scientific revolutions" by analogy with political revolutions. Here the analogy is "turned round," to borrow Frederic Bartlett's phrase (Bartlett, 1932, p. 309), but the reworking of heroes is even more clearly established in science than in official history. As we noted in chapter 4, we agree with critics of Kuhn who point out that scientific thought more often develops gradually than abruptly (M.J. Klein, 1972; Pickering, 1995). Still, Kuhn's more fundamental idea of the creation and modification of scientific paradigms (stable collective memory) from a system of relationships established, and subsequently revised, among observations and theories (labile collective memory) applies to historical as well as scientific collective memory formation, and the analogy extends to the shifting meanings attached to fixed, distinct, items of stable collective memory such as the roles of heroes.

Soviet history textbooks provide a clear example of entity effects on collective consolidation, with the Communist Party exerting an overwhelming influence on the process. The Soviet Communist Party installed its own social hippocampus and directed propaganda bureaucrats to select specific labile items from the collective buffer, to relate them with the appropriate values, and to generalize them into a version of history that emphasized the good of socialism and the evil of capitalism. These ready-made "memories" were imposed upon the masses and influenced the memory consolidation processes of Soviet citizens (D. King, 1997; Schuman & Corning, 2000; P. Jones, 2008). They continue to be felt today (Khazanov, 2008).

The example of Soviet Russia is certainly not offered to suggest that an oppressive regime necessarily forces collective memory formation in a socialist direction. Examples of dictatorships, monarchies, occupying powers, and even decentralized groups within democracies that have attempted to force collective memory formation in other directions can no doubt be found. The larger point is summed up by George Orwell's famous dictum from *1984*: "Who controls the present controls the past" (Orwell, 1949/1961, p. 35). We can recast this dictum in terms of our model: Who controls the present can reshape collective memory by directing the selection of labile items from the collective memory buffer, the choice of meanings to which they are related, and the frequency with which they are presented to the collective generalizer for consolidation into stable collective memory. Thus can collective entity effects on memory formation be abused.

The Continuing Process of Collective Memory Consolidation

Evidence of the influence of the remembering entity on collective memory formation supports our model, but the strongest evidence we present for the existence of the process of collective memory consolidation involves collective retrograde amnesia (see chapters 9–11). The phenomenon of retrograde amnesia occurs because memory consolidation takes time. This is illustrated by what is arguably still the strongest evidence for memory consolidation on the individual level, the finding of retrograde amnesia after damage to the hippocampus and related brain regions. The defining characteristic of this retrograde amnesia is impairment of recent memory (relative to the time of hippocampal damage) but sparing of remote memory (Cohen & Squire, 1981; Rempel-Clower et al., 1996; Bayley et al., 2006). The recent memories that are lost were those in the process of being consolidated (Albert, 1984; Squire, 1992; Cohen & Eichenbaum, 1993). The temporal aspect of individual memory consolidation also occurs on collective levels, as we show in this last section. Two examples of the relatively long time required for collective consolidation concern the collective memories in the United States for the civil rights leader Martin Luther King Jr. and for the Second World War.

President Ronald Reagan signed the Martin Luther King Jr. Federal Holiday bill into law in 1983, but it had been proposed by Representative John Conyers in every legislative session since 1968, the year Martin Luther King Jr. was assassinated (Romero, 2009). This delay of 15 years was caused, not surprisingly, by legislative wrangling—precisely the sort of "contest and negotiation" described by Wertsch (2002) as part of the process of collective memory formation. Similarly, the Martin Luther King Jr. National Memorial Project Foundation spent decades in gaining congressional approval for the construction of a monument to Martin Luther King Jr. on the Mall in Washington, D.C. Much of the delay was bureaucratic, and considerable time was required to raise sufficient funds, but final approval was not won without a certain amount of legislative wrangling of its own (Keefe, 2009). The much less controversial Second World War monument was not even proposed until 1987 (by Representative Marcy Kaptur), was not authorized until 1993, and citing details were not settled until 2001 (Weimer, 2002).

The construction of a monument, like the construction of any stable memory structure, requires time for execution, but low-level operations (e.g., modifying a synapse or cutting a stone) are fast compared with functions that attend the establishment of relationships between a selected memory item and other existing, labile and stable memory items. Reorganization of a system of relationships, especially when it involves politically vested interests, can consume a substantial amount of time. A labile item will not readily be assimilated into stable collective memory if doing so requires a large amount of reorganization. It is possible that, over the course of stable memory formation and reformation, a labile item can readily be assimilated that could not have been at an earlier date. In 1968, the year Martin Luther King Jr. was assassinated, the U.S. Congress was not ready to accept the idea of a federal holiday in his honor, and passage of the legislation required 15 years of contest and negotiation. In contrast, by 1987 the U.S. Congress was more receptive to the idea of a Second World War memorial, and the importance of this event had become well established over the 42 years since the end of the war.

The specific examples of the Martin Luther King Jr. and Second World War memorials demonstrate two reasons, on collective levels, why memory consolidation takes time. First, as in the Martin Luther King Jr. example, integration of a labile item into stable memory requires reorganization of the system of relationships already established in stable memory, and greater amounts of reorganization require longer time periods. To put it bluntly, getting an almost completely white Congress to agree to honor a controversial black public figure required substantial reorganization of social frameworks. Second, as in the Second World War example, the significance of an event, and the need to install it as a stable memory item, may not be appreciated until more time has passed.

Both of these mechanisms probably also occur on the individual level, and it should be possible to design experiments to test whether or not they do. It is already known, for instance, that when individual subjects attempt to remember a list of items, those that can be related to existing schemata are remembered better than those that are irrelevant to them (Johnson, 1970; Bransford & Johnson, 1972; Davidson, 1994). Notably, items that are better remembered are not only those that are consistent with schemata, but also those that are inconsistent with them. It would be interesting to determine whether items that are inconsistent with schemata take longer to consolidate than consistent items, and whether the time required for the consolidation of inconsistent items can be decreased by information provided during the consolidation period that makes them consistent with existing schemata.

The long delay between the occurrence of a significant societal event and its commemoration as a monument or major motion picture has been noted by psychologists James Pennebaker and Becky Banasik (1997). They suggest that delay results, in part, because the people most affected by the event, those in their impressionable teens and early twenties, need a period of decades before they acquire the wherewithal to create a major artwork. They further suggest that people tend to distance themselves psychologically from a traumatic event and need a decades-long recovery period before they can deal with it. They counterbalance these suggestions by noting that some members of society are ready to commemorate an event immediately after it occurs, no matter how traumatic (see also Pennebaker & Gonzales, 2009). This leads them to attribute the delay, mainly, to the time required for consensus to be reached concerning the importance and meaning to society of an event or figure, and in this regard their interpretation is consistent with our model of collective memory consolidation.

The examples of the Martin Luther King Jr. and Second World War memorials illustrate the central role played in the collective consolidation process by the opinion leaders and memory makers in a society. The idea for the Martin Luther King Jr. memorial originated with King's former college fraternity brothers, whereas that for the Second World War memorial originated with veterans' groups. These ideas where taken to Congress, where they were debated for years. The discussions occurred between Members of Congress in consultation with interested constituents, who composed the social hippocampus that strove to relate the proposals for the monuments with other political considerations. A trauma that removed this social structure would have prevented the collective consolidation that finally resulted in authorization of the Martin Luther King Jr. and Second World War monuments. Such social traumas, and their consequent disruptions of the collection consolidation process, have occurred.

The social hippocampus we hypothesize is composed not only of legislators and their influential constituents but also of artists, writers, academics, and members of the general public who participate in the negotiations, deliberations, and decisions that ultimately result in the formation of stable structures that support collective memory and serve as focal points for collective remembering. As it is composed mainly of specialized members of society, the social hippocampus could be especially vulnerable to trauma. Nonspecific traumatic life events, whether physical or emotional, sometimes produce amnesia in individuals through damage to the neurologic hippocampus and related structures (Treadway et al., 1992). Nonspecific social traumas such as war, colonization, or severe economic hardship can suspend the process by which societies form and maintain their histories and cultural identities. The Holocaust has served for decades as a focus for discussions concerning trauma and the effect it has had on the formation of collective memory (e.g., Friedländer, 1992; Young, 1993; LaCapra, 1994, 2001; Bernstein, 2005).

Historian Cathy Caruth is interested in how collective memory for traumatic events can be formed (e.g., Caruth, 1995, 1996). According to Caruth, trauma is at first incomprehensible and can only be placed in the collective narrative after a latent period. This view is consistent with our model of collective memory consolidation, because consolidation is a process that takes time, and the required time could be especially long for a traumatic event that is difficult to reconcile with existing stable memory frameworks. But we extend the view of trauma and collective memory by demonstrating that trauma, in addition to being an item to be remembered, can itself disrupt the process of collective memory formation by causing damage to the social hippocampus. Our demonstration involves memory not *of* the trauma, but of events that occurred *before* the trauma, during the pretrauma period.

The hippocampus is central to the consolidation process in individual brains, and it accordingly takes a central place as the relater in our model. Specifically, the hippocampus is responsible for the establishment of relationships between memory items, both new, labile items and older, more stable items. Its role is critical because the purpose of consolidation is the formation of a usable, accessible, coherent, and meaningful knowledge construct that represents the relationships the hippocampus establishes. Trauma to the hippocampus in an individual brain causes loss of the labile memory items that the hippocampus was in the process of relating, but it spares the already consolidated, stable items. We hypothesize, by analogy, that this pattern of retrograde amnesia should also occur on collective levels as a result of trauma to the social hippocampus. Note that collective retrograde amnesia as we define it differs entirely from the more commonly used term "collective amnesia," which denotes a collective denial rather than a true collective memory loss (i.e., I. Chang, 1997). The case we consider in chapters 9–11, that of the Cultural Revolution perpetrated by the Chinese Communist Party (CCP), is a case of trauma to a social hippocampus, and specifically to the relater of mainland Chinese society, which caused a collective retrograde amnesia.

The comparison between the collective memories of the mainland Chinese and of two Chinese diasporic populations (Táiwān Chinese and Northern Thailand KMT villagers) demonstrates that the damage to the social hippocampus (i.e., to the relaters of the society) during the trauma period interrupted memory consolidation for many non-Communist items from the pretrauma period and so reduced or even erased mainland Chinese memory of such items. For example, mainland Chinese after the Cultural Revolution could all recognize the mythic character of Monkey Sūn, but few knew the religious meaning of his story. Similarly, the Thinking Generation's memory from the pretrauma period of the three writers of the Peking school, which promoted literature based on traditional Chinese culture, was practically nonexistent due to suppression by the CCP. Finally, Zhāng Ailíng and her works, through CCP suppression, were completely erased from the Thinking Generation's memory. However, already consolidated items from the remote past, such as Confucian views on human nature, Daoist views of humans and their position in the universe, and Buddhist views of foreordination, still figured prominently in the Thinking Generation's literature. Since the late 1980s, reduced cultural suppression by the CCP and increased contact with non-mainland Chinese has enabled a substantial recovery of mainland Chinese cultural memory. Books by Zhāng Ailíng are especially popular.

Viewing the Cultural Revolution in hindsight and through the lens of our model, it appears that the CCP under Máo Zédōng specifically targeted the Chinese social hippocampus. It disbanded institutions and imprisoned and even murdered individuals involved in preserving traditional Chinese culture or in advancing the culture in ways deemed contrary to the CCP's cause. Similar targeting of "dissidents" occurred during the "purges" in Soviet Russia and in other Communist countries throughout the past century, but Communist parties are clearly not the sole perpetrators of this practice. Regimes at all points along the political spectrum have caused damage to social hippocampi as a means of derailing consolidation of certain collective memories. Less dramatically, but insidiously, regimes have influenced collective memory formation by manipulating aspects of the consolidation process, such as the contents of the collective buffer, the selection of specific items and the larger ideals with which they are related, and the frequency with which certain associations are made available for collective generalization. By introducing concepts such as collective memory consolidation and the social hippocampus, our model provides a new tool that can be used to understand the actions of historical entities who have sought to influence the formation of collective memory.

We may also turn the analogy "round" (Bartlett, 1932) and suggest that the biological analogue of the relater in social systems may be the neurologic hippocampus. Findings from humans and other animals have led many researchers to conclude that the hippocampus binds, or relates, labile and stable memories (Cohen et al., 1997, 1999). The establishment of relationships among memory items is more easily observed on the social than on the neurologic level, and if our analogy holds, then observations on collective levels could be used to generate hypotheses on the individual level. Does something like the contest and negotiation that occurs during the consolidation of collective memory occur in the brain as well? It should be possible to design experiments to test this. Computational models have already demonstrated how certain forms of learning in the brain could occur through mechanisms that combine cooperation and competition between neurons and neuronal assemblies (e.g., Rumelhart & Zipser, 1985; Kohonen, 1988; Anastasio, 2010). The neural circuitry required for this type of learning requires little more than Hebb's rule ("neurons that fire together wire together") and the excitation and inhibition between neurons that are known to exist in real neural networks (e.g., Purves et al., 2004). It would be of interest to determine if cooperation and competition between neurons and neural assemblies is involved in the establishment of interrelationships mediated by the hippocampus.

Our model accounts both for the possibility that already consolidated memory can recur to influence the consolidation of new memory and for the possibility of the non-mnemonic influence of the entity on consolidation. Such influence can be unequivocally observed on collective levels (see chapters 6, 7, and 8). Could the interests of the remembering entity affect memory consolidation on the individual level? Bartlett long ago envisioned this possibility. He observed that an individual's recollections of stories, pictograms, and other items were incomplete and otherwise distorted and that these inaccuracies often could be related to the occupation, culture, and interests of the remembering individual (Bartlett, 1932). Bartlett's largely anecdotal findings in this area require validation by controlled experimentation, but it seems reasonable to assume that the process of individual memory consolidation can be influenced by the personality of the remembering individual.

The most important difference between our hypothetical social hippocampus and a real neurologic hippocampus is that the latter cannot be replaced or regenerated (at least not at the current level of medical science). An authoritarian regime seeks not only to prevent consolidation of dissident memory items but also to force consolidation of conforming items. In the context of our model, it does this by removing an existing social hippocampus and by replacing it with one of its own. The example of Soviet Russia offered by James Wertsch (2002) (see previous section) and our example of the Chinese Cultural Revolution (see part III) show that this strategy is largely successful in interrupting consolidation of new memory items and in changing the meaning of old memory items, but it is largely unsuccessful in erasing long established cultural memory. Once it removes a social hippocampus, a regime must exert great effort to prevent its regeneration. Cultural groups often succeed in the resumption of a process of collective memory consolidation that was interrupted by social trauma inflicted on it through the actions of an oppressive regime or other cause.

The modern world abounds with examples of cultural groups that are striving to regain their pasts. According to our model, this process first requires the regeneration of a social hippocampus, which would be composed of the opinion leaders in academe, the arts, politics, and other areas of society. But our model, as it stands, does not indicate how a social hippocampus might be regenerated. Perhaps it regrows from a remnant. It could involve exchanges between individuals inside and outside the group regaining its culture. It most likely involves a complex interaction between what survives of the narrative that defines a group (generalized memory) and the desire of the group to maintain its cohesion as a group (entity effects). But whatever the process, it underscores the transcendent difference between individuals and collectives—that individuals definitely die but collectives, held together by their ideals as embodied in their narratives, could endure indefinitely.

Today, somewhere on the Chinese mainland, a young woman reads *Love in a Fallen City* by Zhāng Ailíng (Zhāng & Kingsbury, 2007), and her understanding of her cultural history, and her place in it, changes. As terminology, individual and collective memory consolidation might not sound especially romantic, but through them memory is formed, modified, and maintained. The processes of memory consolidation occurring on individual and collective levels are of fundamental importance. Memory consolidation on the individual level is part of what endows each of us with an identity. Memory consolidation on collective levels is a process that forms constellations of cultural meaning that could continue forever.

References

Ah, Chéng (阿城). (1986). 棋王,树王,孩子王. [*The chess master, The lord of trees, The king of children*] 台北:新帝出版社.

Albert, M. S. (1984). Implications of different patterns of remote memory loss for the concept of consolidation. In H. Weingartner & E. S. Parker (Eds.), *Memory consolidation: Psychobiology of cognition* (pp. 211–230). Hillsdale NJ: Erlbaum.

Alberti, S. J. M. M. (2005). Objects and the museum. *Isis, 96*(4), 559–571.

Alvarez, P., & Squire, L. R. (1994). Memory consolidation and the medial temporal lobe: A simple network model. *Proceedings of the National Academy of Sciences of the United States of America, 91*(15), 7041–7045.

Anastasio, T. J. (2010). *Tutorial on neural systems modeling.* Sunderland, MA: Sinauer Associates.

Anderson, B. (1983). *Imagined communities: Reflections on the origin and spread of nationalism.* London: Verso Editions/NLB.

Ankerberg, J., Weldon, J., & Burroughs, D. (2008). *The facts on Halloween.* Eugene, OR: Harvest House Publishers.

Apple, M. W., & Christian-Smith, L. K. (1991). *The politics of the textbook.* New York: Routledge.

Artizov, A. N. (1993). To suit the views of the leader: The 1936 competition for the [best] textbook on the history of the USSR. *Russian Studies in History, 31*(4), 9–29.

Azer, S. A. (2004). Do recommended textbooks contain adequate information about bile salt transporters for medical students? *Advances in Physiology Education, 28*(1–4), 36–43.

Baddeley, A. (2000). The episodic buffer: A new component of working memory? *Trends in Cognitive Sciences, 4*(11), 417–423.

Baddeley, A., & Hitch, G. (1974). Working memory. In G. H. Bower (Ed.), *The psychology of learning and motivation: Advances in research and theory* (pp. 47–89). New York: Academic Press.

Bahrick, H. P. (1984). Semantic memory content in the permastore: Fifty years of memory for Spanish learned in school. *Journal of Experimental Psychology. General, 113*, 1–29.

Bailey, C. H., & Kandel, E. R. (1993). Structural changes accompanying memory storage. *Annual Review of Physiology, 55*, 397–426.

Baird, D. D. (2006). *Authority on the margin: The informal essays of Virginia Woolf, Natsume Soseki, and Zhōu Zuòrén.* Unpublished doctoral dissertation, University of Oregon.

Bak, P. (1996). *How nature works: The science of self-organized criticaltiy.* New York: Copernicus.

Barthes, R. [1984] (1986). *The rustle of language.* [Bruissement de la langue] (Howard, R., Trans.). New York: Hill & Wang.

Bartlett, F. C. (1932). *Remembering: A study in experimental and social psychology.* Cambridge, UK: Cambridge University Press.

Battelle, J. (2005). *The search: The inside story of how Google and its rivals changed everything.* New York: Portfolio.

Bayley, P. J., Hopkins, R. O., & Squire, L. R. (2006). The fate of old memories after medial temporal lobe damage. *Journal of Neuroscience, 26*(51), 13311–13317.

Beaver, W. C. (1939). *Fundamentals of biology, animal and plant.* St. Louis: C. V. Mosby.

Berg, E. A. (1948). A simple objective technique for measuring flexibility in thinking. *Journal of General Psychology, 39*, 15–22.

Berghahn, V. R., & Schissler, H. (1987). *Perceptions of history: International textbook research on Britain, Germany, and the United States.* New York: Berg.

Bergson, H. [1896] (1913). *Matter and memory.* [Matiere et Memoire] (Paul, N. M., & Palmer, W. S., Trans.). New York: Humanities Press.

Bernstein, R. (2005, May 10). Holocaust memorial opens in Berlin. *New York Times,* p. 6.

Berntsen, D., & Bohn, A. (2009). Cultural life scripts and individual life stories. In P. Boyer & J. V. Wertsch (Eds.), *Memory in mind and culture* (pp. 62–82). Cambridge, UK: Cambridge University Press.

Birth, K. (2006a). The immanent past: Culture and psyche at the juncture of memory and history. *Ethos (Berkeley, Calif.), 34*(2), 169–191.

Birth, K. (2006b). Past times: Temporal structuring of history and memory. *Ethos (Berkeley, Calif.), 34*(2), 192–210.

Blatz, C. W., & Ross, M. (2009). Historical memories. In P. Boyer & J. V. Wertsch (Eds.), *Memory in mind and culture* (pp. 223–237). Cambridge, UK: Cambridge University Press.

Blight, D. W. (2009). The memory boom: Why and why now? In P. Boyer & J. V. Wertsch (Eds.), *Memory in mind and culture* (pp. 238–251). Cambridge, UK: Cambridge University Press.

Bliss, T. V. P., & Lomo, T. (1973). Long-lasting potentiation of synaptic transmission in the dentate area of the anaesthetized rabbit following stimulation of the perforant path. *Journal of Physiology, 232*(2), 331–356.

Bloch, M. (1925). Mémoire collective, tradition et coutume. A propos d'un livre récent. [Collective memory, tradition and custom. In reference to a recent book] *Revue De Synthèse Historique, 40*(14), 73–83.

Bloch, M. [1953] (1964). *The historian's craft.* [Apologie pour l'histoire] (Putnám, P., Trans.). New York: Vintage Books.

Boden, M. A. (1990). *The creative mind: Myths and mechanisms.* London: Weidenfeld & Nicolson.

Borges, J. L. [1942] (1962). Funes the Memorious. In D. A. Yates, & J. E. Irby (Eds.), *Labyrinths: Selected stories & other writings* (Aug. ed., pp. 59–66). New York: New Directions Publishing Corporation.

Bourdieu, P. [1972] (1977). *Outline of a theory of practice.* [Esquisse d'une théorie de la pratique] (Nice, R., Trans.). Cambridge, MA: Cambridge University Press.

Bouwsma, W. J. (1990). *A usable past: Essays in European cultural history.* Berkeley: University of California Press.

Bowler, P. J. (1989). *The Mendelian revolution: The emergence of hereditarian concepts in modern science and society.* Baltimore: Johns Hopkins University Press.

Bowman, J. (2006). Seeing what's missing in memories of Cyprus. *Peace Review, 18*(1), 119–127.

Boyer, P. (2001). *Religion explained: The evolutionary origins of religions thought.* New York: Basic Books.

Bracey, C. A. (2008). *Saviors or sellouts: The promise and peril of black conservatism, from Booker T. Washington to Condoleezza Rice.* Boston: Beacon Press.

Brainerd, C. J., & Dempster, F. N. (1995). *Interference and inhibition in cognition.* San Diego: Academic Press.

Brainerd, C. J., & Reyna, V. F. (2001). Fuzzy-trace theory: Dual processes in memory, reasoning, and cognitive neuroscience. *Advances in Child Development and Behavior*, *28*, 41–100.

Brainerd, C. J., & Reyna, V. F. (2005). *The science of false memory*. New York: Oxford University Press.

Bransford, J. D., & Johnson, M. K. (1972). Contextual prerequisites for understanding: Some investigations of comprehension and recall. *Journal of Verbal Learning and Verbal Behavior*, *11*, 717–726.

Brewer, W. F. (1986). What is autobiographical memory? In D. Rubin (Ed.), *Autobiographical memory* (pp. 25–49). Cambridge, UK: Cambridge University Press.

Brewer, W. F. (1987). Schemas versus mental models in human memory. In P. Morris (Ed.), *Modelling cognition* (pp. 187–197). Chichester, UK: Wiley.

Brewer, W. F., & Nakamura, G. V. (1984). The nature and functions of schemas. In R. S. Wyer & T. K. Srull (Eds.), *Handbook of social cognition* (pp. 119–160). Hillsdale, NJ: Erlbaum.

Brewer, W. F., & Pani, J. R. (1983). The structure of human memory. In G. H. Bower (Ed.), *The psychology of learning and motivation: Advances in research and theory* (pp. 1–38). New York: Academic Press.

Brewer, W. F., & Treyens, J. C. (1981). Role of schemata in memory for places. *Cognitive Psychology*, *13*(2), 207–230.

Brown, R., & Kulik, J. (1977). Flashbulb memories. *Cognition*, *5*, 73–99.

Bruner, J. S. (1986). *Actual minds, possible worlds*. Cambridge, MA: Harvard University Press.

Bruner, J. S. (1990). *Acts of meaning*. Cambridge, MA: Harvard University Press.

Bunsey, M., & Eichenbaum, H. (1993). Critical role of the parahippocampal region for paired-associate learning in rats. *Behavioral Neuroscience*, *107*(5), 740–747.

Burnham, W. H. (1903). Retroactive amnesia: Illustrative cases and a tentative explanation. *American Journal of Psychology*, *14*, 382–396.

Burton, A. M. (2003). *Dwelling in the archive: Women writing house, home, and history in late colonial India*. New York: Oxford University Press.

Burton, A. M. (2005). *Archive stories: Facts, fictions, and the writing of history*. Durham, NC: Duke University Press.

Butler, A. B., & Hodos, W. (1996). *Comparative vertebrate neuroanatomy: Evolution and adaptation*. New York: John Wiley & Sons, Inc.

Butters, N., Miliotis, P., Albert, M. S., & Sax, D. S. (1984). Memory assessment: Evidence of heterogeneity of amnesic symptoms. In G. Goldstein (Ed.), *Advances in clinical neuropsychology* (pp. 127–159). New York: Plenum Press.

Camerer, C., Loewenstein, G., & Prelec, D. (2005). Neuroeconomics: How neuroscience can inform economics. *Journal of Economic Literature, 43*(1), 9–64.

Canguilhem, G. [1966] (1978). *On the normal and the pathological.* [Le normal et le pathologique] (Fawcett, C. R., Trans.). Boston: D. Reidel.

Carey, B. (2008, December 5). H. M., an unforgettable amnesiac, dies at 82. *New York Times,* p. A1.

Caruth, C. (1995). *Trauma: Explorations in memory.* Baltimore: Johns Hopkins University Press.

Caruth, C. (1996). *Unclaimed experience: Trauma, narrative, and history.* Baltimore: Johns Hopkins University Press.

Celizic, M. (2007, June 11). "Baby Jessica" 20 years later. *TODAY.* Retrieved from http://www.msnbc.msn.com/id/19165433/.

Chan, N. F. (2004). *The "Yigu" movement and the construction of "Baihua" literary history: A case study of Hú shì and Gu Jiegang.* Unpublished doctoral dissertation, Hong Kong University of Science and Technology.

Chan, W. (1978). *Religious trends in modern China.* New York: Octagon Books.

Chang, I. (1997). *The Rape of Nanking: The forgotten holocaust of World War II.* New York: Basic Books.

Chāng, Wen-Chin (1999). *Beyond the military: The complex migration and resettlement of the KMT Yunnanese Chinese in northern Thailand.* Unpublished doctoral dissertation, Katholieke Universiteit Leuven, Belgium.

Chāng, Wen-Chin (2001). From war refugees to immigrants: The case of the KMT Yunnanese Chinese in Northern Thailand. *International Migration Review, 35,* 1086–1105.

Chāng, Wen-Chin (2002). Identification of leadership among the KMT Yunnanese Chinese in Northern Thailand. *Journal of Southeast Asian Studies, 33,* 123–146.

Chāng, Wen-Chin (2003). Three Yunnanese jade traders from Tengchong. *Kolor: Journal on Moving Communities, 3,* 15–34.

Chāng, Wen-Chin (2004). Guanxi and regulation in networks: The Yunnanese jade trade between Burma and Thailand, 1962–88. *Journal of Southeast Asian Studies, 35*(3), 479–501.

Chāng, Wen-Chin (2006). Home away from home: Migrant Yunnanese Chinese in Northern Thailand. *International Journal of Asian Studies, 1,* 49–76.

Chén, Fāngmíng (陈芳明). (1999). 张爱玲与台湾文学史的书写. [Zhāng Ailíng and the writing of a history of Chinese literature in Táiwān] In Yáng Zé (杨泽) (Ed.), 《阅读张爱玲: 张爱玲国际研讨会论文集》[Reading Zhāng Ailíng: A paper collection from the International Conference on Zhāng Ailíng] 台湾:麦田出版股份有限公司.

Chen, Kenneth K. S. (1964). *Buddhism in China: A historical survey*. Princeton: Princeton University Press.

Chén, Sīhé (陈思和). (1999). 民间与现代都市文化:兼论张爱玲现象. [Folk and modern urban culture: Also on the phenomenon of Zhāng Ailíng] In Yáng Zé (杨泽) (Ed.), 《阅读张爱玲:张爱玲国际研讨会论文集》[Reading Zhāng Ailíng: Collection of papers from the International Conference on Zhāng Ailíng] 台湾: 麦田出版股份有限公司.

Chén, Yashu. (1998). *Love demythologized: The significance and impact of Zhāng Ailíng's (1921–1995) works*. Unpublished doctoral dissertation, University of Wisconsin-Madison.

Chén, Zǐshàn (陈子善). (2003). 张爱玲的风气:1949年前的张爱玲评论 [The vogue of Zhāng Ailíng: Commentaries on Zhāng Ailíng before 1949] 济南:山东书画出版社.

Chéng, C. (1977). The impact of Japanese literary trends. In M. Goldman (Ed.), *Modern Chinese literature in the May Fourth Era* (pp. 63–87). Cambridge, MA: Harvard University Press.

Chéng, Maorong. (1999). *Literary modernity: Studies in Lu Xun and Shen Congwen*. Unpublished doctoral dissertation, University of British Columbia.

Chinn, C. A., & Brewer, W. F. (1996). Mental models in data interpretation. *Philosophy of Science, 63*(s1), S221–S219.

Chomsky, N. (1965). *Aspects of the theory of syntax*. Cambridge, MA: MIT Press.

Chow, K. (1994). *The rise of Confucian ritualism in late imperial China: Ethics, classics, and lineage discourse*. Stanford: Stanford University Press.

Chun, M. M., & Turk-Browne, N. B. (2007). Interactions between attention and memory. *Current Opinion in Neurobiology, 17*(2), 177–184.

Clayton, N. S., Griffiths, D. P., Emery, N. J., & Dickinson, A. (2001). Elements of episodic-like memory in animals. *Philosophical Transactions of the Royal Society B. Biological Sciences, 356*(1413), 1491.

Cohen, N. J., & Eichenbaum, H. (1993). *Memory, amnesia, and the hippocampal system*. Cambridge, MA: MIT Press.

Cohen, N. J., & Squire, L. R. (1980). Preserved learning and retention of pattern-analyzing skill in amnesia: Dissociation of knowing how and knowing that. *Science, 210*(4466), 207–210.

Cohen, N. J., & Squire, L. R. (1981). Retrograde amnesia and remote memory impairment. *Neuropsychologia, 19*(3), 337–356.

Cohen, N. J., Eichenbaum, H., & Poldrack, R. A. (1997). Memory for items and memory for relations in the procedural/declarative memory framework. *Memory (Hove, England), 5*(1–2), 131–178.

Cohen, N. J., Ryan, J., Hunt, C., Romine, L., Wszalek, T., & Nash, C. (1999). Hippocampal system and declarative (relational) memory: Summarizing the data from functional neuroimaging studies. *Hippocampus, 9*(1), 83–98.

Cohen, R. (1997). *Global diasporas: An introduction.* London: UCL Press.

Cohn, M. A., Mehl, M. R., & Pennebaker, J. W. (2004). Linguistic markers of psychological change surrounding September 11, 2001. *Psychological Science, 15,* 687–693.

Collins, H. M., & Evans, R. (2002). The third wave of science studies: Studies of expertise and experience. *Social Studies of Science, 32*(2), 235–296.

Commager, H. S. (1967). *The search for a usable past, and other essays in historiography.* New York: Knopf.

Confino, A. (1997). Collective memory and cultural history: Problems of method. *American Historical Review, 102*(5), 1386.

Conway, M. A., & Pleydell-Pearce, C. W. (2000). The construction of autobiographical memories in the self memory system. *Psychological Review, 107,* 261–288.

Conway, M. A., Wang, Q., Hanyu, K., & Haque, S. (2005). A cross-cultural investigation of autobiographical memory—on the universality and cultural variation of the reminiscence bump. *Journal of Cross-Cultural Psychology, 36,* 739–749.

Corkin, S. (1984). Lasting consequences of bilateral medial temporal lobectomy: Clinical course and experimental findings in H.M. *Seminars in Neurology, 4,* 249–259.

Corkin, S. (2002). What's new with the amnesic patient H.M.? *Nature Reviews. Neuroscience, 3*(2), 153–160.

Cowan, N. (2001). The magical number 4 in short-term memory: A reconsideration of mental storage capacity. *Behavioral and Brain Sciences, 24*(1), 87.

Cowan, N. (2005). *Working memory capacity.* New York: Psychology Press.

Crane, S. A. (1997). Writing the individual back into collective memory. *American Historical Review, 102*(5), 1372–1385.

Crew, S. R. (1996). Who owns history?: History in the museum. *History Teacher, 30*(1), 83–88.

Csikszentmihalyi, M. (1996). *Creativity: Flow and the psychology of discovery and invention*. New York: HarperCollins.

Curci, A., & Luminet, O. (2006). Follow-up of a cross-national comparison on flashbulb and event memory for the September 11th attacks. *Memory (Hove, England), 14,* 329–344.

Curci, A., Luminet, O., Finkenauer, C., & Gisle, L. (2001). Flashbulb memories in social groups: A comparative test–retest study of the memory of French President Mitterrand's death in a French and a Belgian group. *Memory (Hove, England), 9*(2), 81–101.

Daruvala, S. (2000). *Zhou Zuoren and an alternative Chinese response to modernity*. Cambridge, MA: Harvard University Asia Center.

Davidson, D. (1994). Recognition and recall of irrelevant and interruptive atypical actions in script-based stories. *Journal of Memory and Language, 33,* 757–775.

De Bruyn, P. (2004). Wǔdāng shan: The origin of a major center of modern Daoism. In J. Lagerwey (Ed.), *Religion and Chinese society* (pp. 553–590). Hong Kong: Chinese University Press & Ecole Française d'Extrème-Orient.

Deese, J. (1959). On the prediction of occurrence of particular verbal intrusions in immediate recall. *Journal of Experimental Psychology, 58,* 17–22.

DelFattore, J. (1992). *What Johnny shouldn't read: Textbook censorship in America*. New Haven, CT: Yale University Press.

De Rivera, J. (1998). Relinquishing believed-in imaginings: Narratives of people who have repudiated false accusations. In J. de Rivera & T. R. Sarbin (Eds.), *Believed-in imaginings: The narrative construction of reality* (pp. 169–188). Washington, DC: American Psychological Association.

De Rivera, J., & Sarbin, T. R. (1998). *Believed-in imaginings: The narrative construction of reality*. Washington, DC: American Psychological Association.

Derrida, J. [1995] (1996). *Archive fever: A Freudian impression*. [Mal d'archive] (Prenowitz, E., Trans.). Chicago: University of Chicago Press.

Dillon, M., & Minority Rights Group. (2001). *Religious minorities and China*. London: Minority Rights Group International.

Dobson, R. B. (Ed.). (1970). *The Peasants' Revolt of 1381*. New York: St Martin's Press.

Draaisma, D. (2000). *Metaphors of memory: A history of ideas about the mind*. New York: Cambridge University Press.

Duàn, Yĭng (段颖). (2002). 边陲侨乡的历史、记忆与象征:云南腾冲和顺宗族、社会变迁的个案研究 [History, memory, and symbolization in a frontier oversees Chinese village: A case study of lineage and social vicissitude in Hénshùn County, Téngchōng Area, Yúnnán Province] In 载陈志明、丁毓玲、王连茂主编≪跨国网络与华南侨—文

化、认同和社会变迁≫ [Transnational networks and overseas Chinese in south China] 香港中文大学, 香港亚太研究所出版.

Duara, P. (1995). *Rescuing history from the nation: Questioning narratives of modern China*. Chicago: University of Chicago Press.

Dubois, T. (2004). Village community and reconstruction of religious life in rural north China. In J. Lagerwey (Ed.), *Religion and Chinese society* (pp. 837–868). Hong Kong: Chinese University Press & Ecole Française d'Extrème-Orient.

Du Bois, W. E. B. (1935). *Black reconstruction*. New York: Russell & Russell.

Duke, M. S. (1985). *Blooming and contending: Chinese literature in the post-Mao era*. Bloomington, IN: Indiana University Press.

Dunbar, K. (1997). How scientists think: Online creativity and conceptual change in science. In T. B. Ward, S. M. Smith, & J. Vaid (Eds.), *Creative thought: An investigation of conceptual processes and structures* (pp. 461–493). Washington, DC: American Psychological Association.

Duncan, C. P. (1949). The retroactive effect of electroshock on learning. *Journal of Comparative and Physiological Psychology, 42*, 32–44.

Durkheim, E. [1893] (1997). *The division of labor in society*. [De la division du travail social] (Halls, W. D., Trans.). New York: Free Press.

Eakin, P. J. (1999). *How our lives become stories: Making selves*. Ithaca, NY: Cornell University Press.

Ebbinghaus, H. (1885). *Über das Gedächtnis: Untersuchungen zur experimentellen Psychologie*. [On memory: Investigations in experimental psychology] Leipzig: Duncker & Humbolt.

Eichenbaum, H. (2000). A cortical-hippocampal system for declarative memory. *Nature Reviews. Neuroscience, 1*(1), 41–50.

Eichenbaum, H. (2003a). How does the hippocampus contribute to memory? *Trends in Cognitive Sciences, 7*(10), 427–429.

Eichenbaum, H. (2003b). Learning and memory: Brain systems. In L. R. Squire, F. E. Bloom, S. K. McConnell, J. L. Roberts, N. C. Spitzer, & M. J. Zigmond (Eds.), *Fundamental neuroscience* (2nd ed., pp. 1299–1326). San Diego: Academic Press.

Eichenbaum, H. (2006). Memory binding in hippocampal relational networks. In H. D. Zimmer, A. Mecklinger, & U. Lindenberger (Eds.), *Handbook of binding and memory: Perspectives from cognitive neuroscience* (pp. 25–51). New York: Oxford University Press.

Eichenbaum, H., & Otto, T. (1993). Odor-guided learning and memory in rats: Is it 'special'? *Trends in Neurosciences, 16*(1), 22–24, discussion 25–26.

Ellen, R. F. (1986). What Black Elk left unsaid: On the illusory images of green primitivism. *Anthropology Today, 2*(6), 8–12.

Elson, R. M. (1964). *Guardians of tradition, American schoolbooks of the nineteenth century.* Lincoln, NE: University of Nebraska Press.

Engelhardt, T., & Linenthal, E. T. (1996). *History wars: The Enola Gay and other battles for the American past.* New York: Metropolitan Books/Henry Holt & Co.

Evans, A., Jr. (1976). Trends in Soviet secondary school histories of the USSR. *Soviet Studies, 28*(2), 224–243.

Feldman, D. H., Gardner, H., & Csikszentmihalyi, M. (1994). *Changing the world: A framework for the study of creativity.* Westport, CT: Praeger.

Finkenauer, C., Luminet, O., Gisle, L., El-Ahmadi, A., Van Der Linden, M., & Philippot, P. (1998). Flashbulb memories and the underlying mechanism of their formation: Toward an emotional-integrative model. *Memory & Cognition, 26*, 516–531.

FitzGerald, F. (1979). *America revised: History schoolbooks in the twentieth century.* Boston: Little, Brown & Co.

Fleck, L. [1935] (1979). T. J. Trenn & R. K. Merton (Eds.), *Genesis and development of a scientific fact.* [Entstehung und Entwicklung einer wissenschaftlichen Tatsache] (Bradley, F., & Trenn, T. J., Trans.). Chicago: University of Chicago Press.

Forster, E. M. (1927). *Aspects of the novel.* New York: Harcourt, Brace & Co.

Foster, J. K., & Wilson, A. C. (2005). Sleep and memory: Definitions, terminology, models, and predictions? *Behavioral and Brain Sciences, 28*(1), 71.

Foucault, M. [1966] (1971). *The order of things: An archaeology of the human sciences* [Mots et les choses] (Unknown Trans.). New York: Pantheon Books

Foucault, M. [1969] (1972). *The archaeology of knowledge.* [Archéologie du savoir] (Smith, A. M. S., Trans.). New York: Pantheon Books.

Frank, M. J., Loughry, B., & O'Reilly, R. C. (2001). Interactions between frontal cortex and basal ganglia in working memory: A computational model. *Cognitive, Affective & Behavioral Neuroscience, 1*(2), 137–160.

Franklin, J. H., & Meier, A. (1982). *Black leaders of the twentieth century.* Urbana, IL: University of Illinois Press.

Freud, S. [1900] (1953). J. Strachey, C. L. Rothgeb, & A. Freud (Eds.), *The standard edition of the complete psychological works of Sigmund Freud.* London: Hogarth Press.

Freud, S. [1913] (1952). *Totem and taboo. Some points of agreement between the mental lives of savages and neurotics.* [Totem und Tabu] (Strachey, J., Trans.). New York: Norton.

Freud, S. [1937] (1939). *Moses and monotheism.* [Der Mann Moses und die monotheistische Religion] (Jones, K., Trans.). London: Hogarth Press and the Institute of Psycho-Analysis.

Friedländer, S. (1992). *Probing the limits of representation: Nazism and the "final solution"*. Cambridge, MA: Harvard University Press.

Frisch, M. (1989). American history and the structures of collective memory: A modest exercise in empirical iconography. *Journal of American History, 75*(4), 1130–1155.

Fù, Jiāqín (傅家勤). (1984). 中国道教史 [The history of Daoism in China] 上海: 上海书店.

Fung, A. S. (1997). Accuracy of current educational literature on the staging of gastric carcinoma. *World Journal of Surgery, 21*(3), 237–239.

Fuster, J. M., Bauer, R. H., & Jervey, J. P. (1985). Functional interactions between inferotemporal and prefrontal cortex in a cognitive task. *Brain Research, 30*, 299–307.

Gallagher, E. J. (2000). *The Enola Gay controversy—about—overview*. Retrieved from http://digital.lib.lehigh.edu/trial/enola/about/.

Galtung, J., & Ruge, M. (1965). The structure of foreign news: The presentation of the Congo, Cuba and Cyprus crises in four Norwegian newspapers. *Journal of Peace Research, 2*(1), 64–90.

Gardner, H. (1982). *Art, mind, and brain: A cognitive approach to creativity*. New York: Basic Books.

Gardner, H. (1993). *Creating minds: An anatomy of creativity seen through the lives of Freud, Einstein, Picasso, Stravinsky, Eliot, Graham, and Gandhi*. New York: Basic Books.

Gaster, B. (1990a). Assimilation of scientific change: The introduction of molecular genetics into biology textbooks. *Social Studies of Science, 20*(3), 431–454.

Gaster, B. (1990b). The slow diffusion of the DNA paradigm into biology textbooks. *Trends in Biochemical Sciences, 15*(8), 325–327.

Gates, H. L., & West, C. (2000). *The African-American century: How black Americans have shaped our country*. New York: Free Press.

Gazzaniga, M. S. (1997). *Conversations in the cognitive neurosciences*. Cambridge, MA: MIT Press.

Geary, P. J. (2002). *Myth of nations: The medieval origins of Europe*. Princeton, NJ: Princeton University Press.

Gedi, N., & Elam, Y. (1996). Collective memory—what is it? *History & Memory, 8*(1), 30–50.

Gentner, D., & Loftus, E. F. (1979). Integration of verbal and visual information as evidenced by distortions in picture memory. *American Journal of Psychology, 92*(2), 363–375.

Gerbner, G. (1973). Cultural indicators: The third voice. In G. Gerbner, L. P. Gross, & W. H. Melody (Eds.), *Communications technology and social policy: Understanding the new 'cultural revolution'* (pp. 555–573). New York: Wiley.

Gnadt, J. W., & Andersen, R. A. (1988). Memory related motor planning activity in posterior parietal cortex of macaque. *Experimental Brain Research, 70,* 216–220.

Grieder, J. B. (1970). *Hu Shih and the Chinese renaissance: Liberalism in the Chinese Revolution, 1917–1937.* Cambridge, MA: Harvard University Press.

Griffiths, D., Dickinson, A., & Clayton, N. (1999). Episodic memory: What can animals remember about their past? *Trends in Cognitive Sciences, 3*(2), 74–80.

Gulick, J. T. (1905). *Evolution, racial and habitual.* Washington, DC: Carnegie Institute.

Gunn, Edward M. (1980). *Unwelcome Muse: Chinese Literature in Shànghǎi and Peking 1937-45.* New York: Columbia Univeristy Press.

Habermas, J. [1962] (1989). *The structural transformation of the public sphere.* [Strukturwandel der Öffentlichkeit: Untersuchungen zu einer Kategorie der bürgerlichen Gesellschaft] (Berger, T., & Lawrence, F., Trans.). Cambridge, MA: MIT Press.

Halberstam, J. (2005). *In a queer time and place: Transgender bodies, subcultural lives.* New York: New York University Press.

Halbwachs, M. (1925) [1992]. *Les cadres sociaux de la mémoire. [The social frameworks of memory].* Paris: Félix Alcan.

Halbwachs, M. [1950] (1980). J. Alexandre (Ed.), *The collective memory.* [Mémorie collective] (Ditter, F. J., Jr., & Ditter, W. Y., Trans.). New York: Harper & Row.

Halbwachs, M., & Coser, L. A. (Eds.). [1925] (1992). *On collective memory.* Chicago: University of Chicago Press.

Hall, B. K. (2006a). "Evolutionist and missionary," the Reverend John Thomas Gulick (1832–1923). Part I: Cumulative segregation—geographical isolation. *Journal of Experimental Zoology. Part B. Molecular and Developmental Evolution, 306B*(5), 407–418.

Hall, B. K. (2006b). "Evolutionist and missionary," the Reverend John Thomas Gulick (1832–1923). Part II: Coincident or ontogenetic selection—the Baldwin effect. *Journal of Experimental Zoology. Part B. Molecular and Developmental Evolution, 306B*(6), 489–495.

Hanks, W. F. (1996). *Language and communicative practices.* Boulder, CO: Westview Press.

Hannula, D. E., Federmeier, K. D., & Cohen, N. J. (2006a). Event-related potential signatures of relational memory. *Journal of Cognitive Neuroscience, 18,* 1863–1876.

Hannula, D. E., Tranel, D., & Cohen, N. J. (2006b). The long and the short of it: Relational memory impairments in amnesia, even at short lags. *Journal of Neuroscience, 26,* 8352–8359.

Harcup, T., & O'Neill, D. (2001). What is news? Galtung and Ruge revisited. *Journalism Studies, 2*(2), 261–280.

Hartz, P. (1993). *Taoism.* New York: Facts on File.

Hayes, J. R. M. (1952). *Memory span for several vocabularies as a function of vocabulary size.* Cambridge, MA: Acoustics Laboratory, Massachusetts Institute of Technology.

Hazy, T. E., Frank, M. J., & O'Reilly, R. C. (2006). Banishing the homunculus: Making working memory work. *Neuroscience, 139*(1), 105–118.

Head, H. (in conjunction with W. H. R. Rivers, G. Holmes, J. Sherren, T. Thompson, & G. Riddoch) (1920). *Studies in neurology.* London: Oxford University Press.

Hebb, D. O. (1949). *The organization of behavior: A neuropsychological theory.* New York: Wiley.

Hebb, D. O. (1972). *Textbook of psychology* (3rd ed.). Toronto: W. B. Saunders.

Helfand, W. H. (1990). Art in the service of public health: The illustrated poster. *Caduceus: Art, Science and Politics in the Service of Public Health, 6*(2), 1–37.

Helfand, W. H., & Keister, L. H. (1990). *To your health: An exhibition of posters for contemporary public health issues.* Bethesda, MD: National Library of Medicine, U.S. Department of Health and Human Services, Public Health Service, National Institutes of Health.

Henig, R. M. (2000). *The monk in the garden: The lost and found genius of Gregor Mendel, the father of genetics.* Boston: Houghton Mifflin.

Herrmann, D. J., & Chaffin, R. (1988). *Memory in historical perspective: The literature before Ebbinghaus.* New York: Springer-Verlag.

Hirsch, A. (1995, June 28). Enola Gay exhibit opens without an agenda. *Baltimore Sun,* p. 1D.

Ho, A. K., Milchan, A., & Stone, O. (Producers) & Stone, O. (Director). (1991). *JFK.* [Motion Picture] United States: Le Studio Canal+, Regency Enterprises, Alcor Films & Ixtlan Corporation.

Hobsbawm, E. J. (1962). *The age of revolution, 1789–1848.* New York: New American Library.

Hobsbawm, E. J. (1975). *The age of capital, 1848–1875.* London: Weidenfeld and Nicolson.

Hobsbawm, E. J., & Ranger, T. O. (1983). *The invention of tradition.* New York: Cambridge University Press.

Hoddeson, L. (2007). In the wake of Thomas Kuhn's theory of scientific revolutions: The perspective of an historian of science. In S. Vosniadou, A. Baltas, & X. Vamvakoussi (Eds.), *Reframing the conceptual change approach in learning and instruction* (pp. 25–34). Boston: Elsevier.

Holmes, F. L. (2001). *Meselson, Stahl, and the replication of DNA: A history of "the most beautiful experiment in biology"*. New Haven, CT: Yale University Press.

Hoobler, T., & Hoobler, D. (1993). *Confucianism*. New York: Facts on File.

Hopfield, J. J. (1982). Neural networks and physical systems with emergent collective computational abilities. *Proceedings of the National Academy of Sciences of the United States of America, 79*(8), 2554–2558.

Howe, M. L., Reyna, V. F., & Brainerd, C. J. (1992). *Development of long-term retention*. New York: Springer-Verlag.

Hoyan, C. H. F. (1996). *The life and works of Zhāng Ailíng: A critical study*. Unpublished doctoral dissertation, University of British Columbia.

Hsia, C. (1999). *A history of modern Chinese fiction* (3rd ed.). Bloomington, IN: Indiana University Press.

Huáng, Jiàn (黃健). (2002). 京派文学与批评研究. [Literature and critique of the Peking School] 上海:三联书店.

Hunter, J. (2002). Minds, archives, and the domestication of knowledge. In R. Comay (Ed.), *Lost in the archives* (pp. 199–217). Toronto: Alphabet City Media.

Huntley, W. L. (2006). Rebels without a cause: North Korea, Iran, and the NPT. *International Affairs, 82*, 723–742.

Hutton, P. H. (1993). *History as an art of memory*. Burlington, VT: University of Vermont.

Hutton, P. H. (1994). Sigmund Freud and Maurice Halbwachs: The problem of memory in historical psychology. *History Teacher, 27*(2), 145–158.

Hyman, I. E., Jr., & Loftus, E. F. (1998). Errors in autobiographical memory. *Clinical Psychology Review, 18*(8), 933–947.

James, W. (1890). *The principles of psychology*. New York: Henry Holt & Co.

James, W. [1899] (1983). *Talks to teachers on psychology and to students on some of life's ideals*. Cambridge, MA: Harvard University Press.

Jì, X. (1999). 在胡适墓前的回忆和反思. [Recollection and reflection in front of Hú Shì's tomb] Retrieved from http://news.ifeng.com/history/1/jishi/200907/0713_2663_1246532.shtml.

Jiāng, C. (江燦滕). (1996). 台湾佛教百年史之研究. [A study of the century of Buddhism in Táiwān] 南天书局有限公司.

Jiāng, C. (江燦滕). (2003). 战后台湾传统佛教的教派发展与现代社会. [The development of traditional Buddhist sects and modern society after World War II] 载《台湾本土宗教研究的新视野与新思维》, 张珣, 江燦藤主编, 南天书局有限公司.

Johnson, R. E. (1970). Recall of prose as a function of the structural importance of the linguistic units. *Journal of Verbal Learning and Verbal Behavior, 9*, 12–20.

Jonas, G. (2005). *Freedom's sword: The NAACP and the struggle against racism in America, 1909–1969*. New York: Routledge.

Jones, C. B. (1999). *Buddhism in Taiwan: Religion and the state, 1660–1990*. Honolulu, HI: University of Hawaii's Press.

Jones, P. (2008). Memories of terror or terrorizing memories? Terror, trauma and survival in Soviet culture of the thaw. *Slavonic and East European Review, 86*(2), 346–371.

Kansteiner, W. (2002). Finding meaning in memory: A methodological critique of collective memory studies. *History and Theory, 41*(2), 179–197.

Kansteiner, W. (2007). Alternate worlds and invented communities: History and historical consciousness in the age of interactive media. In K. Jenkins, S. Morgan, & A. Munslow (Eds.), *Manifestos for history* (pp. 131–148). New York: Routledge.

Kavanagh, G. (1996). *Making histories in museums*. New York: Leicester University Press.

Kē, Líng [1984] (1991). 遥寄张爱玲 [Memoir of Zhāng Ailíng from a distance]. In Y. Qing & J. Hongda (Eds.), 张爱玲研究资料 [Materials for the study of Zhāng Ailíng]. 海峡文艺出版社.

Keefe, B (2009) MLK memorial tied up in red tape. *The Atlanta Journal-Constitution*, August 25 edition.

Keister, L. H. (1990). The poster collection at the national library of medicine. *Caduceus: Art, Science and Politics in the Service of Public Health, 6*(2), 38–42.

Kennedy, P. M. (1987). *The rise of the great powers*. New York: Random House.

Khazanov, A. M. (2008). Whom to mourn and whom to forget? (Re)constructing collective memory in contemporary Russia. *Totalitarian Movements & Political Religions, 9*(2), 293–310.

Kieschnick, J. (2003). *The impact of Buddhism on Chinese material culture*. Princeton, NJ: Princeton University Press.

King, A. Y. C. (1996). State Confucianism and its transformation: The restructuring of the state-society relation in Táiwān. In W. Tu (Ed.), *Confucian traditions in East Asian modernity: Moral education and economic culture in Japan and the four mini-dragons* (pp. 228–243). Cambridge, MA: Harvard University Press.

King, D. (1997). *The commissar vanishes: The falsification of photographs and art in Stalin's Russia*. New York: Metropolitan Books.

Kinkley, J. C. (1985). *After Mao: Chinese literature and society 1978–1981*. Cambridge, MA: Harvard University Press.

Kirkpatrick, E. A. (1894). An experimental study of memory. *Psychological Review, 1*, 602–609.

Klein, K. L. (2000). On the emergence of memory in historical discourse. *Representations, 69* (Special Issue: Grounds for Remembering), 127–150.

Klein, M. J. (1972). The use and abuse of historical teaching in physics. In S. G. Brush & A. L. King (Eds.), *History in the teaching of physics* (pp. 12–18). Hanover, NH: University Press of New England.

Knudsen, E. I. (2007). Fundamental components of attention. *Annual Review of Neuroscience, 30*, 57–78.

Koch, W. (2007, May 29). Lives of indelible impact. *USA Today*, p. 8B.

Kohlstedt, S. G. (2005). "Thoughts in things": Modernity, history, and North American museums. *Isis, 96*(4), 586–601.

Kohonen, T. (1988). *Self-organization and associative memory* (2nd ed.). New York: Springer-Verlag.

Krige, J. (1980). *Science, revolution, and discontinuity*. Atlantic Highlands, NJ: Humanities Press.

Kronick, D. A. (2004). *"Devant le deluge" and other essays on early modern scientific communication*. Lanham, MD: Scarecrow Press.

Kuhn, T. S. [1962] (1996). *The structure of scientific revolutions* (3rd ed.). Chicago: University of Chicago Press.

Kutler, S. I. (1990). *The wars of Watergate: The last crisis of Richard Nixon*. New York: Knopf.

LaCapra, D. (1994). *Representing the Holocaust: History, theory, trauma*. Ithaca, NY: Cornell University Press.

LaCapra, D. (2001). *Writing history, writing trauma*. Baltimore: Johns Hopkins University Press.

Lambert, A. J., Scherer, L. N., Rogers, C., & Jocoby, L. (2009). How does collective memory create a sense of the collective? In P. Boyer & J. V. Wertsch (Eds.), *Memory in mind and culture* (pp. 194–217). Cambridge, UK: Cambridge University Press.

Lashley, K. (1950). In search of the engram. In J. F. Danielli & R. Brown (Eds.), *Physiological mechanisms in animal behavior* (pp. 454–482). Cambridge, UK: Cambridge University Press.

Latour, B. (1987). *Science in action: How to follow scientists and engineers through society.* Philadelphia: Open University Press.

Latour, B., & Woolgar, S. (1979). *Laboratory life: The social construction of scientific facts.* Beverly Hills, CA: Sage Publications.

Le Bon, G. [1895] (1960). *The crowd; a study of the popular-mind.* [La Psychologie des foules] (Merton, R. K., Trans.). New York: Viking Press.

LeCun, Y., Boser, B., Denker, J. S., Henderson, D., Howard, R. E., Hubbard, W., et al. (1989). Backpropagation applied to handwritten zip code recognition. *Neural Computation, 1,* 541–551.

Lee, L. O. (1973). *The romantic generations of modern Chinese writers.* Cambridge, MA: Harvard University Press.

L'Engle, M. (1962). *A wrinkle in time.* New York: Farrar, Straus, & Giroux.

Lewis, D. J., Misanin, J. R., & Miller, R. R. (1968). Recovery of memory following amnesia. *Nature, 220*(5168), 704–705.

Li, M. (1990). *Hushi and his Deweyan reconstruction of Chinese history.* Unpublished doctoral dissertation, Boston University.

Li Qingxi (2000). Searching for roots: Anticultural return in mainland Chinese literature of the 1980s. In D. D-W. Wang & P-Y. Chi (Eds.), *Chinese literature in the second half of a modern century: A critical survey.* Bloomington and Indianapolis: Indiana University Press.

Lǐ, S. (李尚全). (2006). 当代中国汉传佛教信仰方式的变迁 [The transformation of worship formats of contemporary Chinese Mahayana Buddhism in China] 甘肃人民出版社.

Lín, Cháochéng (林朝成). (2003). 台湾人间佛教与环境论述 [A discussion on the relation between the worldly Buddhism of Táiwān and the environment] 载台湾本土宗教研究的新视野与新思维 [New approaches to the study of native religions in Táiwān], 张珣,江燦藤主编,南天书局有限公司.

Lindberg, D. C., & Westman, R. S. (1990). *Reappraisals of the scientific revolution.* New York: Cambridge University Press.

Líng, Yu (凌宇). (1991). 沈从文概论 [A general discussion on Shěn Cóngwén]. In Shao. Huaqiang (Ed.), 沈从文研究资料 [Research Materials for studying Shěn Cóngwén]. Huacheng Chubanshe.

Link, E. P. (1983). *Stubborn weeds: Popular and controversial Chinese literature after the Cultural Revolution.* Bloomington, IN: Indiana University Press.

Little, S., & Eichman, S. (2000). *Realm of the immortals: Daoism in the arts of China: The Cleveland Museum of Art, February 10–April 10, 1988* [Qian jing]. Cleveland, OH: Cleveland Museum of Art & Indiana University Press.

Littlefield, D. F. (1992). American Indians, American scholars and the American literary canon. *American Studies (Lawrence, Kan.)*, *33*, 95–111.

Loftus, E. F. (1996a). The myth of repressed memory and the realities of science. *Clinical Psychology: Science and Practice*, *3*, 356–362.

Loftus, E. F. (1996b). Memory distortion and false memory creation. *Bulletin of the American Academy of Psychiatry and the Law*, *24*, 281–295.

Loftus, E. F. (1997). Creating false memories. *Scientific American*, *277*(3), 70–75.

Loftus, E. F., & Ketcham, K. (1994). *The myth of repressed memory: False memories and allegations of sexual abuse*. New York: St. Martin's Press.

Loftus, E. F., & Pickrell, J. E. (1995). The formation of false memories. *Psychiatric Annals*, *25*, 720–725.

Lomo, T. (2003). The discovery of long-term potentiation. *Philosophical Transactions of the Royal Society of London. Series B, Biological Sciences*, *358*(1432), 617–620.

Louie, K. (1985). Love stories: The meaning of love and marriage in China. In J. C. Kinkley (Ed.), *After Mao: Chinese literature and society, 1978–1981* (pp. 67–86). Cambridge, MA: Council on East Asian Studies, Harvard University.

Lupu, N. (2003). Memory vanished, absent, and confined. *History & Memory*, *15*(2), 130–164.

Lynn, S. J., & McConkey, K. M. (1998). *Truth in memory*. New York: Guilford Press.

Malkki, L. H. (1995). *Purity and exile: Violence, memory, and national cosmology among Hutu refugees in Tanzania*. Chicago: University of Chicago Press.

Manns, J. R., Hopkins, R. O., Reed, J. M., Kitchener, E. G., & Squire, L. R. (2003). Recognition memory and the human hippocampus. *Neuron*, *37*(1), 171–180.

Markowitsch, H. J., & Pritzel, M. (1985). The neuropathology of amnesia. *Progress in Neurobiology*, *25*(3), 189–287.

Marx, K., & Engels, F. (1970). *The German ideology* [Deutsche Ideologie] (C. J. Arthur, Ed.). New York: International Publishers. (Originally written 1846, published 1932.)

Mavratsas, C. V. (1997). The ideological contest between Greek-Cypriot nationalism and cypriotism 1974–1995: Politics, social memory and identity. *Ethnic and Racial Studies*, *20*(4), 717–737.

McClelland, J. L., & Rogers, T. T. (2003). The parallel distributed processing approach to semantic cognition. *Nature Reviews. Neuroscience*, *4*, 310–322.

McClelland, J. L., McNaughton, B. L., & O'Reilly, R. C. (1995). Why there are complementary learning systems in the hippocampus and neocortex: Insights from the successes and failures of connectionist models of learning and memory. *Psychological Review, 102*(3), 419–457.

McCloskey, M., & Cohen, N. J. (1989). Catastrophic interference in connectionist networks: The sequential learning problem. *Psychology of Learning and Motivation, 24*, 109–165.

McCloskey, M., Wible, C. G., & Cohen, N. J. (1988). Is there a special flashbulb-memory mechanism? *Journal of Experimental Psychology. General, 117*(2), 171–181.

McDonald, R. J., & White, N. M. (1993). A triple dissociation of memory systems: Hippocampus, amygdala, and dorsal striatum. *Behavioral Neuroscience, 107*(1), 3–22.

McDougall, B. S. (1977). The impact of Western literary trends. In M. Goldman (Ed.), *Modern Chinese literature in the May Fourth Era* (pp. 37–62). Cambridge, MA: Harvard University Press.

McDougall, B. S., & Mao, Z. [1943] (1980). *Mao Zedong's "Talks at the Yan'an Conference on Literature and Art": A translation of the 1943 text with commentary.* Ann Arbor, MI: Center for Chinese Studies, University of Michigan.

McDougall, W. (1901). Experimentelle Beitraege zur Lehre vom Gedaechtnis: Von G.E. Mueller und A. Pilzecker [Experimental contributions to the teaching of memory: by G.E. Mueller and A. Pilzecker] *Mind, 10*, 388–394.

McGaugh, J. L. (1972). Impairment and facilitation of memory consolidation. *Activitas Nervosa Superior, 14*(1), 64–74.

McGaugh, J. L. (2000). Memory—a century of consolidation. *Science, 287*(5451), 248–251.

McGaugh, J. L., & Herz, M. J. (1972). *Memory consolidation.* San Francisco, CA: Albion.

McGaugh, J. L., Gold, P. E., Handwerker, M. J., Jensen, R. A., & Martinez, J. (1979). Altering memory by electrical and chemical stimulation of the brain. In M. A. B. Brazier (Ed.), *Brain mechanisms in memory and learning: From the single neuron to man* (pp. 151–164). New York: Raven.

McIntyre, J. W. (1998). Evolution of 20th-century attitudes to prophylaxis of pulmonary aspiration during anaesthesia. *Canadian Journal of Anaesthesia / Journal Canadien d'Anesthesie, 45*(10), 1024–1030.

McLeod, P. J. (1991). How to produce instructional text for a medical audience. *Medical Teacher, 13*(2), 135–144.

Mehl, M. R., & Pennebaker, J. W. (2003). The social dynamics of a cultural upheaval: Social interactions surrounding September 11, 2003. *Psychological Science, 14*, 579–585.

Meisner, M. J. (1999). *Mao's China and after: A history of the People's Republic* (3rd ed.). New York: Free Press.

Metzger, T. A. (1977). *Escape from predicament: Neo-confucianism and China's evolving political culture*. New York: Columbia University Press.

Miller, G. A. (1956). The magical number seven, plus or minus two: Some limits on our capacity for processing information. *Psychological Review, 101*(2), 343–352.

Milner, P. M. (1999). *The autonomous brain: A neural theory of attention and learning*. Mahwah, NJ: L. Erlbaum Associates.

Milner, B., Corkin, S., & Teuber, H. L. (1968). Further analysis of the hippocampal amnesia syndrome: 14 year follow-up study of H. M. *Neuropsychologia, 6*, 215–234.

Misanin, J. R., Miller, R. R., & Lewis, D. J. (1968). Retrograde amnesia produced by electroconvulsive shock after reactivation of a consolidated memory trace. *Science, 160*(827), 554–555.

Mishkin, M. (1978). Memory in monkeys severely impaired by combined but not separate removal of the amygdala and hippocampus. *Nature, 273*, 297–298.

Mishra, P., & Brewer, W. F. (2003). Theories as a form of mental representation and their role in the recall of text information. *Contemporary Educational Psychology, 28*(3), 277–303.

Moll, J., Zahn, R., de Oliveira-Souza, R., Krueger, F., & Grafman, J. (2005). The neural basis of human moral cognition. *Nature Reviews. Neuroscience, 6*(10), 799–809.

Moreau, J. (2003). *Schoolbook nation: Conflicts over American history textbooks from the Civil War to the present*. Ann Arbor, MI: University of Michigan Press.

Müller, G. E., & Pilzecker, A. (1900). Experimentelle Beiträge zur Lehre vom Gedächtnis. *Zeitschrift für Psychologie. Ergänzungsband, 1*, 1–300.

Murray, D. J. (1968). Articulation and acoustic confusability in short-term memory. *Journal of Experimental Psychology, 78*(4), 679–684.

Musil, R. [1936] (1987). *Posthumous papers of a living author*. [Nachlass zu Lebzeiten] (Wortsman, P., Trans.). Hygiene, CO: Eridanos Press.

Nadel, L., Ryan, L., Hayes, S. M., Gilboa, A., & Moscovitch, M. (2003). The role of the hippocampal complex in long-term episodic memory. *International Congress Series, 1250*, 215–234.

Nader, K. (2003). Memory traces unbound. *Trends in Neurosciences, 26*(2), 65–72.

Nader, K., Schafe, G. E., & LeDoux, J. E. (2000). The labile nature of consolidation theory. *Nature Reviews. Neuroscience, 1*(3), 216–219.

Nalbantian, S. (2003). *Memory in literature: From Rousseau to neuroscience*. New York: Palgrave Macmillan.

Neisser, U. (1967). *Cognitive psychology*. New York: Appleton-Century-Crofts.

Neisser, U. (1981). John Dean's memory: a case study. *Cognition, 9*, 102–115.

Neisser, U. (1988). Five kinds of self-knowledge. *Philosophical Psychology, 1*(1), 35–59.

Neisser, U., & Harsch, N. (1992). Phantom flashbulbs: False recollections of hearing the news about Challenger. In E. Winograd & U. Neisser (Eds.), *Affect and accuracy in recall: Studies of "flashbulb memories"* (pp. 9–31). New York: Cambridge University Press.

Nesmith, T. (2002). Seeing archives: Postmodernism and the changing intellectual place of archives. *American Archivist, 65*, 24–41.

Ng, M. (1988). *The Russian hero in modern Chinese fiction*. Hong Kong: Chinese University Press; New York: State University of New York Press.

Nivison, D. S. (1956). Communist ethics and Chinese tradition. *Journal of Asian Studies, 16*(1), 51–74.

Nora, P. (1984). *Les lieux de mémoire*. [Places of memory] Paris: Gallimard.

Nora, P. (1989). Between memory and history: *Les lieux de memoire*. *Representations (Berkeley, Calif.), 26*(1), 7–24.

Norman, D. A., & Shallice, T. (1986). Attention to action: Willed and automatic control of behaviour. In D. E. Schwartz & D. Shapiro (Eds.), *Consciousness and self-regulation* (pp. 376–390). New York: Plenum Press.

Novick, P. (1999). *The Holocaust in American life*. Boston: Houghton Mifflin.

O'Keefe, J., & Nadel, L. (1978). *The hippocampus as a cognitive map*. New York: Oxford University Press.

Olick, J. K. (1998). Introduction to memory and the nation. *Social Science History, 22*(4), 377.

Olick, J. K. (1999). Collective memory: The two cultures. *Sociological Theory, 17*(3), 333–348.

Olick, J. K., & Robbins, J. (1998). Social memory studies: From "collective memory" to the historical sociology of mnemonic practices. *Annual Review of Sociology, 24*(1), 105–140.

Olney, J. (1998). *Memory and narrative: The weave of life-writing*. Chicago: University of Chicago Press.

Orwell, G. [1949] (1961). *Nineteen eighty-four, a novel*. New York: Penguin.

Otis, L. (1994). *Organic memory: History and the body in the late nineteenth and early twentieth centuries*. Lincoln, NE: University of Nebraska Press.

Otis, L. (1999). *Membranes: Metaphors of invasion in nineteenth-century literature, science, and politics*. Baltimore: Johns Hopkins University Press.

Otis, L. (2001). *Networking: Communicating with bodies and machines in the nineteenth century*. Ann Arbor, MI: University of Michigan Press.

Otis, L. (2002). The metaphoric circuit: Organic and technological communication in the nineteenth century. *Journal of the History of Ideas, 63*(1), 105–128.

Pan, Y., & Pan, J. (1985). The non-official magazine *Today* and the younger generation's ideals for a new literature. In J. C. Kinkley (Ed.), *After Mao: Chinese literature and society, 1978–1981* (pp. 193–220). Cambridge, MA: Council on East Asian Studies, Harvard University.

Papadakis, Y. (1994). The National Struggle Museums of a divided city. *Ethnic and Racial Studies, 17*(3), 400–419.

Parker, K., & Dean, C. (1997). *Ten years of the Pew News Interest Index*. Washington, DC: The Pew Research Center for the People and the Press.

Peking University. (1979). 中国现代文学史 [A history of Chinese modern literature]. In J. X. Zhāng & Z. Tierong (Eds.), 周作人研究资料 [Complied materials for the study of Zhōu Zuòrén], 天津人民出版社.

Pendergrast, M. (1996). *Victims of memory*. Hinesberg, VT: Upper Access Books.

Pennebaker, J. W., & Banasik, B. L. (1997). On the creation and maintenance of collective memories: History as social psychology. In J. W. Pennebaker, D. Paez, & B. Rimé (Eds.), *Collective memory of political events: Social psychological perspectives* (pp. 3–19). Mahwah, NJ: Lawrence Erlbaum Associates.

Pennebaker, J. W., & Gonzales, A. L. (2009). Making history: Social and psychological processes underlying collective memory. In P. Boyer & J. V. Wertsch (Eds.), *Memory in mind and culture* (pp. 171–193). Cambridge, UK: Cambridge University Press.

Pennebaker, J. W., Rimé, B., & Páez, D. (1997). *Collective memory of political events: Social psychological perspectives*. Mahwah, NJ: Lawrence Erlbaum Associates.

Pickering, A. (1995). *The mangle of practice: Time, agency, and science*. Chicago: University of Chicago Press.

Pinsker, H., Kupfermann, I., Castellucci, V., & Kandel, E. (1970). Habituation and dishabituation of the gill-withdrawal reflex in *Aplysia*. *Science, 167*, 1740–1742.

Pitman, R. K., Sanders, K. M., Zusman, R. M., Healy, A. R., Cheema, F., Lasko, N. B., et al. (2002). Pilot study of secondary prevention of posttraumatic stress disorder with propranolol. *Biological Psychiatry, 51*(2), 189–192.

Poa, D. (1992). *The discourse on human nature in twentieth century Chinese literature*. Unpublished doctoral dissertation, Stanford University.

Purves, D., Augustine, G. J., Fitzpatrick, D., Hall, W. C., LaMantia, A.-S., McNamara, J. O., & Williams, S. M. (Eds.). (2004). *Neuroscience: Third Edition*. Sunderland, MA: Sinauer.

Qián, L. (钱理群). (1991). 周作人论. [On Zhōu Zuòrén] 上海人民出版社.

Quillian, M. R. (1968). Semantic memory. In M. L. Minsky (Ed.), *Semantic information processing* (pp. 227–270). Cambridge, MA: MIT Press.

Quillian, M. R. (1969). The teachable language comprehender. *Communications of the Association for Computing Machinery, 12*, 459–475.

Quinn, N. (Ed.). (2005). *Finding culture in talk: A collection of methods*. New York: Palgrave Macmillan.

Rempel-Clower, N. L., Zola, S. M., Squire, L. R., & Amaral, D. G. (1996). Three cases of enduring memory impairment after bilateral damage limited to the hippocampal formation. *Journal of Neuroscience, 16*(16), 5233–5255.

Ribot, T. [1875] (1973). *Heredity: A psychological study of its phenomena, laws, causes, and consequences*. New York: D. Appleton & Co.

Ribot, T. (1881). *Les maladies de la memoire*. [Illnesses of memory]. Paris: Germer Bailliere.

Riccio, D. C., Millin, P. M., & Bogart, A. R. (2006). Reconsolidation: A brief history, a retrieval view, and some recent issues. *Learning & Memory (Cold Spring Harbor, N.Y.), 13*(5), 536–544.

Richardson, R., Riccio, D. C., & Mowrey, H. (1982). Retrograde amnesia for previously acquired pavlovian conditioning: UCS exposure as a reactivation treatment. *Physiological Psychology, 10*, 384–390.

Roediger, H. L., & Goff, L. M. (1998). Memory. In W. Bechtel, G. Graham, & D. A. Balota (Eds.), *A companion to cognitive science* (pp. 250–264). Malden, MA: Basil Blackwell.

Roediger, H. L., & McDermott, K. B. (1995). Creating false memories: Remembering words not presented in lists. *Journal of Experimental Psychology. Learning, Memory, and Cognition, 21*, 803–814.

Roediger, H. L., Zaromb, F. M., & Butler, A. C. (2009). The role of repeated retrieval in shaping collective memory. In P. Boyer & J. V. Wertsch (Eds.), *Memory in mind and culture* (pp. 138–170). Cambridge, UK: Cambridge University Press.

Rolls, E. T. (2007). An attractor network in the hippocampus: Theory and neurophysiology. *Learning & Memory (Cold Spring Harbor, N.Y.), 14*, 714–731.

Rolls, E. T., & Kesner, R. P. (2006). A computational theory of hippocampal function, and empirical tests of the theory. *Progress in Neurobiology, 79*(1), 1–48.

Romanes, G. J. (1892–1897). In C. L. Morgan (Ed.), *Darwin and after Darwin. An exposition of the Darwinian thought and a discussion of post-Darwinian questions* (Vols. 2 & 3). Chicago: Open Court.

Romero, F. (2009). A brief history of Martin Luther King Jr. Day. *Time, 173*(2). Retrieved from http://www.time.com/time/nation/article/0,8599,1872501,00.html.

Rosenberg, E. S. (2003). *A date which will live: Pearl Harbor in American memory.* Durham, NC: Duke University Press.

Rosenblatt, F. [1958] (1983). *The perceptron.* Buffalo, NY: Cornell Aeronautical Laboratory.

Rosenblatt, F. (1958). The perceptron: A probabilistic model for information storage and organization in the brain. *Psychological Review, 65,* 386–408.

Rosenthal, D. B. (1985). Evolution in high school biology textbooks: 1963–1983. *Science Education, 69,* 637–648.

Ross, M., & Wilson, A. E. (2003). Autobiographical memory and conceptions of self: Getting better all the time. *Current Directions in Psychological Science, 12,* 66–69.

Rossington, M., & Whitehead, A. (Eds.). (2007). *Theories of memory: A reader.* Edinburgh: Edinburgh University Press.

Rothberg, M. (2009). *Multidirectional memory: Remembering the Holocaust in the age of decolonization.* Stanford, CA: Stanford University Press.

Rozin, P., Millman, L., & Nemeroff, C. (1986). Operation of the laws of sympathetic magic in disgust and other domains. *Journal of Personality and Social Psychology, 50,* 703–712.

Rubin, D. C., & Berntsen, D. (2003). Life scripts help to maintain autobiographical memories of highly positive, but not highly negative, events. *Memory & Cognition, 31,* 1–14.

Rubin, D. C., Wetzler, S. E., & Nebes, R. D. (1986). Autobiographical memory across the adult lifespan. In D. C. Rubin (Ed.), *Autobiographical memory* (pp. 202–221). New York: Cambridge University Press.

Rumelhart, D. E., & McClelland, J. L. (1986). *Parallel distributed processing: Explorations in the microstructure of cognition.* Cambridge, MA: MIT Press.

Rumelhart, D. E., & Todd, P. M. (1993). Learning and connectionist representations. In D. E. Meyer & S. Kornblum (Eds.), *Attention and performance XIV: Synergies in experimental psychology, artificial intelligence, and cognitive neuroscience* (pp. 3–30). Caimbridge, MA: MIT Press.

Rumelhart, D. E., & Zipser, D. (1985). Feature discovery by competitive learning. *Cognitive Science, 9*(1), 75–112.

Sacks, O. W. [1985] (1998). *The man who mistook his wife for a hat and other clinical tales.* New York: Simon & Schuster.

Sahdra, B., & Ross, M. (2007). Group identification and historical memory. *Personality and Social Psychology Bulletin, 33,* 384–395.

Sarbin, T. R. (1986). *Narrative psychology: The storied nature of human conduct.* New York: Praeger.

Schacter, D. L. (1987). Implicit memory: History and current status. *Journal of Experimental Psychology. Learning, Memory, and Cognition, 13,* 501–518.

Schacter, D. L. (Ed.). (1995). *Memory distortion: How minds, brains, and societies reconstruct the past.* Cambridge, MA: Harvard University Press.

Schacter, D. L. (1996). *Searching for memory: The brain, the mind, and the past.* New York: BasicBooks.

Schacter, D. L., & Scarry, E. (Eds.). (2001). *Memory, brain, and belief.* Cambridge, MA: Harvard University Press.

Schacter, D. L., & Tulving, E. (1994). *Memory systems.* Cambridge, MA: MIT Press.

Schacter, D. L., Addis, D. R., & Buckner, R. L. (2007). Remembering the past to imagine the future: The prospective brain. *Nature Reviews. Neuroscience, 8,* 657–661.

Schiff, S. (2010). *Cleopatra: A life.* New York: Little, Brown.

Schroeder, J. (2003). *Stars and Stripes* sold out. *Reminisce, 13*(6), 30–31.

Schudson, M. (1992). *Watergate in American memory: How we remember, forget, and reconstruct the past.* New York: Basic Books.

Schudson, M. (1994). Textbook politics. *Journal of Communication, 44*(1), 43–51.

Schuman, H., & Corning, A. D. (2000). Collective knowledge of public events: The Soviet era from the great purge to glasnost. *American Journal of Sociology, 105*(4), 913.

Schuman, H., Schwartz, B., & D'Arcy, H. (2005). Elite revisionists and popular beliefs: Christopher Columbus, hero or villain? *Public Opinion Quarterly, 69,* 2–29.

Schwartz, B. (1998). Postmodernity and historical reputation: Abraham Lincoln in late twentieth-century American memory. *Social Forces, 77,* 63–102.

Schwartz, B. (2000). *Abraham Lincoln and the forge of national memory.* Chicago: University of Chicago Press.

Schwartz, B. (2003). Lincoln at the millennium. *Journal of the Abraham Lincoln Association, 24*(1), 1–31.

Schwartz, B. (2008). *Abraham Lincoln in the post-heroic era: History and memory in late twentieth-century America.* Chicago: University of Chicago Press.

Scott, J. (2002). Mapping the past: Turkish Cypriot narratives of time and place in the Canbulat Museum, Northern Cyprus. *History & Anthropology, 13*(3), 217–230.

Scoville, W. B., & Milner, B. (1957). Loss of recent memory after bilateral hippocampal lesions. *Journal of Neurology, Neurosurgery, and Psychiatry, 20*, 11–21.

Seremetakis, C. N. (Ed.). (1996). *The senses still: Perception and memory as material culture in modernity*. Chicago: University of Chicago Press.

Shallice, T. (1988). *From neuropsychology to mental structure*. New York: Cambridge University Press.

Shannon, C. E., & Weaver, W. W. (1949). *The mathematical theory of communication*. Urbana: University of Illinois Press.

Shohat, E. (2006). *Taboo memories, diasporic voices*. Durham, NC: Duke University Press.

Siu, H. F., & Stern, Z. (1983). *Mao's harvest: Voices from China's new generation*. New York: Oxford University Press.

Skoog, G. (1984). The coverage of evolution in high school biology textbooks published in the 1980s. *Science Education, 68*, 117–128.

Smith, R. (2003). From publication to change. *BMJ (Clinical Research Ed.), 327*(7405).

Smith, E. R., & DeCoster, J. (2000). Dual-process models in social and cognitive psychology: Conceptual integration and links to underlying memory systems. *Personality and Social Psychology Review, 4*(2), 108–131.

Snow, C. P. (1959). *The two cultures and the scientific revolution*. New York: Cambridge University Press.

Spickard, P. R., & Burroughs, W. J. (2000). *We are a people: Narrative and multiplicity in constructing ethnic identity*. Philadelphia: Temple University Press.

Squire, L. R. (1992). Memory and the hippocampus: A synthesis from findings with rats, monkeys, and humans. *Psychological Review, 99*(2), 195–231.

Squire, L. R., & Alvarez, P. (1995). Retrograde amnesia and memory consolidation: A neurobiological perspective. *Current Opinion in Neurobiology, 5*(2), 169–177.

Squire, L. R., & Bayley, P. J. (2007). The neuroscience of remote memory. *Current Opinion in Neurobiology, 17*(2), 185–196.

Squire, L. R., & Cohen, N. J. (1979). Memory and amnesia: Resistance to disruption develops for years after learning. *Behavioral and Neural Biology, 25*(1), 115–125.

Squire, L. R., & Zola, S. M. (1996). Structure and function of declarative and nondeclarative memory systems. *Proceedings of the National Academy of Sciences of the United States of America, 93*(2419), 13515–13522.

Squire, L. R., Cohen, N. J., & Zouzounis, J. A. (1984). Preserved memory in retrograde amnesia: Sparing of a recently acquired skill. *Neuropsychologia, 22*(2), 145–152.

Stone, L. D., & Pennebaker, J. W. (2002). Trauma in real time: Talking and avoiding online conversations about the death of Princess Diana. *Basic and Applied Social Psychology, 24*, 172–182.

Stroop, J. R. (1935). Studies of interference in serial verbal reactions. *Journal of Experimental Psychology, 18*(6), 643–662.

Stoler, A. L. (2002). Colonial archives and the arts of governance. *Archival Science, 2*, 87–109.

Stuckey, G. A. (2005). *Memory–tradition–history: Ties to the past in modern Chinese fiction.* Unpublished doctoral dissertation, University of California at Los Angeles.

Tacke, C. (2000). *1848: Memory and oblivion in Europe.* New York: P.I.E.-Peter Lang.

Tai, H. H. (2001). Remembered realms: Pierre Nora and French national memory. *American Historical Review, 106*(3), 906–922.

Tanaka, K., Saito, H., Fukada, Y., & Moriya, M. (1991). Coding visual images of objects in the inferotemporal cortex of the macaque monkey. *Journal of Neurophysiology, 66*(1), 170–189.

Táng, Dàcháo (唐大潮). (2001). 中国道教简史. [A brief history of Daoism in China] 宗教文化出版社.

Táng, Dégāng (唐德刚). (1991). 胡适的历史地位与历史作用. [Hú Shì's position and role in history] In《胡适与近代中国》[Hú Shì and modern China] 时报出版社.

Tián, Z. (田仲济), & Sūn, C. (孙昌熙). (1979). 中国现代文学史. [A history of modern Chinese literature] 山东人民出版社.

Treadway, M., McCloskey, M., Gordon, B., & Cohen, N. J. (1992). Landmark life events and the organization of memory: Evidence from functional retrograde amnesia. In S.-A. Christianson (Ed.), *The handbook of emotion and memory: Research and theory* (pp. 389–410). Hillsdale, NJ: Lawrence Erlbaum Associates.

Treisman, A. (1996). The binding problem. *Current Opinion in Neurobiology, 6*, 171–178.

Trezise, T. (2001). Unspeakable. *Yale Journal of Criticism, 14*(1), 39–66.

Trinkler, I., King, J. A., Spiers, H. J., & Burgess, N. (2006). Part or parcel? Contextual binding of events in episodic memory. In H. D. Zimmer, A. Mecklinger, & U. Lindenberger (Eds.), *Handbook of binding and memory: Perspectives from cognitive neuroscience* (pp. 53–83). New York: Oxford University Press.

Tulving, E. (1972). Episodic and semantic memory. In E. Tulving & W. Donaldson (Eds.), *Organization of memory* (pp. 381–403). London: Academic Press.

Tulving, E. (1983). *Elements of episodic memory*. New York: Oxford University Press.

Tyack, D. B. (1999). Monuments between covers: The politics of textbooks. *American Behavioral Scientist, 42*(6), 922–932.

Tyack, D. B. (2003). *Seeking common ground: Public schools in a diverse society.* Cambridge, MA: Harvard University Press.

Vähä, F. M. E. (2002). Producing patriots: Heroic stories and individual heroes as the makers of the Soviet and Russian identity in history textbooks (1950–1995). *Ab Imperio [Russia]*, (3), 545–559.

Vasterman, P. L. M. (2005). Media-hype: Self-reinforcing news waves, journalistic standards and the construction of social problems. *European Journal of Communication, 20*(4), 508–530.

Wah, W. Y. (1988). *Essays on Chinese literature: A comparative approach*. Kent Ridge, Singapore: Singapore University Press.

Waldum, C., Zhao, C. M., & Chen, D. (2001). Are current textbooks good enough for physiology education? For example, the ECL cells are missing. *Advances in Physiology Education, 25*(1–4), 123–126.

Wallace, A. R. (1889). *Darwinism: An exposition of the theory of natural selection. With some of its applications.* New York: Macmillan.

Wáng, A. (2008). *The song of everlasting sorrow: A novel of Shanghai. Translated by Michael Berry and Susan Chan Egan.* Columbia University Press.

Wáng, Déwēi (王德威). (2001). 落地的麦子不死: 张爱玲的文学影响力与"张派"作家的超越之路 [The wheat that falls on the ground will not die: The influence of Zhāng Ailíng in literature and the persistence of "the school of Zhāng"] In Zǐ Tóng and Yī Qīng (Eds.), ≪张爱玲评说六十年≫[Sixty years of commentary on and evaluation of Zhāng Ailíng] 中国华侨出版社.

Wang, D. D.-W., & Chi P.-Y., eds. (2000). *Chinese literature in the second half of a modern century: A critical survey.* Bloomington: Indiana University Press.

Wáng, Y. [1951] (1986). A draft history of Chinese New Literature. In J. X. Zhāng & Z. Tiěróng (Eds.), *Complied materials for the study of Zhōu Zuòrén*, 天津人民出版社。.

Wáng, Yáo (王瑶). [1954] (1991). 中国新文学史稿 [Draft of a history of the new Chinese literature], In H. Shao (Ed.), 沈从文研究资料 [Research materials for studying Shěn Cóngwén].花城出版社

Wáng, Z. (王子德). (1989). 序言. [Introduction] In ≪胡适研究丛录≫ [Materials for scholarship on Hú Shì] 三联书店.

Wang, S. H., Ostlund, S. B., Nader, K., & Balleine, B. W. (2005). Consolidtion and reconsolidation of incentive learning in the amygdala. *Journal of Neuroscience, 25*(4), 830–835.

Ward, K. (2006). *History in the making: An absorbing look at how American history has changed in the telling over the last 200 years.* New York: The New Press.

Watson, J. B. (1930). *Behaviorism* (rev. ed.). Chicago: University of Chicago Press.

Watson, J. D., & Crick, F. H. (1953). Molecular structure of nucleic acids. A structure for deoxyribose nucleic acid. *Nature, 171*(4356), 737–738.

Wegner, D. M. (1986). Transactive memory: A contemporary analysis of the group mind. In B. Mullen & G. R. Goethals (Eds.), *Theories of group behavior* (pp. 185–205). New York: Springer-Verlag.

Weisberg, R. W. (1993). *Creativity: Beyond the myth of genius.* New York: W. H. Freeman.

Weimer, D. R. (2002). Legislative history of the World War II memorial and World War II commemorative legislation. Congressional Research Service, The Library of Congress, Report for Congress, Order Code RL31390.

Welch, H. (1968). *The Buddhist revival in China.* Cambridge, MA: Harvard University Press. '

Welch, H. (1972). *Buddhism under Mao.* Cambridge, MA: Harvard University Press.

Wellman, B. (Ed.). (1999). *Networks in the Global Village.* Boulder, CO: Westview Press.

Wertsch, J. V. (1991). *Voices of the mind: A sociocultural approach to mediated action.* Cambridge, MA: Harvard University Press.

Wertsch, J. V. (1998). *Mind as action.* New York: Oxford University Press.

Wertsch, J. V. (2002). *Voices of collective remembering.* New York: Cambridge University Press.

Wertsch, J. V. (2009). Collective memory. In P. Boyer & J. V. Wertsch (Eds.), *Memory in mind and culture* (pp. 117–137). Cambridge, UK: Cambridge University Press.

Wertsch, J. V., & Roediger, H. L. (2008). Collective memory: Conceptual foundations and theoretical approaches. *Memory (Hove, England), 16*(3), 318.

White, G. (2006). Epilogue: Memory moments. *Ethos (Berkeley, Calif.), 34*(2), 325–341.

White, H. V. (1973). *Metahistory: The historical imagination in nineteenth-century Europe.* Baltimore: Johns Hopkins University Press.

White, H. V. (1980). The value of narrativity in the representation of reality. *Critical Inquiry, 7*(1), 5–27.

White, H. V. (1987). *The content of the form: Narrative discourse and historical representation.* Baltimore: Johns Hopkins University Press.

Whitehouse, H. (2004). *Modes of religiosity: A cognitive theory of religions transmission.* Walnut Creek, CA: Altamira Press.

Wilford, J. N. (1991). *The mysterious history of Columbus: An exploration of the man, the myth, the legacy.* New York: Knopf.

Williams, H. L., & Conway, M. A. (2009). Networks of autobiographical memories. In P. Boyer & J. V. Wertsch (Eds.), *Memory in mind and culture* (pp. 3–28). Cambridge, UK: Cambridge University Press.

Wilson, A. E., & Ross, J. (2001). From chump to champ: People's appraisals of their earlier and current selves. *Journal of Personality and Social Psychology, 80,* 572–584.

Winerip, M. (1998, January 11). Looking for an 11 o'clock fix. *New York Times,* p. SM30.

Winstanley, M. (1976). Assimilation into the literature of a critical advance in molecular biology. *Social Studies of Science, 6*(3/4; Special Issue: Aspects of the Sociology of Science: Papers from a Conference, University of York, UK, 16–18 September 1975), 545–549.

Winter, J. (2009). Historians as sites of memory. In P. Boyer & J. V. Wertsch (Eds.), *Memory in mind and culture* (pp. 252–268). Cambridge, UK: Cambridge University Press.

Winter, J. M., & Sivan, E. (Eds.). (1999). *War and remembrance in the twentieth century.* New York: Cambridge University Press.

Woodell, H. (1993). *All the king's men: The search for a usable past.* New York: Twayne.

Woodward, B. (1999). *Shadow: Five presidents and the legacy of Watergate.* New York: Simon & Schuster.

Wyatt, J. (1991). Use and sources of medical knowledge. *Lancet, 338*(8779), 1368–1373.

Xiao, J. (2004). *Memory and women in modern Chinese literature: Shen Congwen, Zhang Ailing, and Wang Anyi.* Unpublished doctoral dissertation, The State University of New Jersey.

Yáng, H. (2007–2008). Personal correspondence with W. Zhang.

Yáng, Z. (杨泽), Ed. (1999). ≪阅读张爱玲:张爱玲国际研讨会论文集≫[Reading Zhāng Ailíng: A collection of papers from the International Conference on Zhāng Ailíng] 台湾:麦田出版股份有限公司.

Yates, F. A. (1966). *The art of memory.* Chicago: University of Chicago Press.

Yerushalmi, Y. H. (1982). *Zakhor, Jewish history and Jewish memory.* Seattle: University of Washington Press.

Ying, L. (1992). *Answering the call of tradition: The root-seeking movement of contemporary Chinese literature*. Unpublished doctoral dissertation, University of Texas at Austin.

Yonelinas, A. P. (2002). The nature of recollection and familiarity: A review of 30 years of research. *Journal of Memory and Language, 46*(3), 441–517.

Young, J. E. (1993). *The texture of memory: Holocaust memorials and meaning*. New Haven, CT: Yale University Press.

Young, J. E. (2000). *At memory's edge: After-images of the Holocaust in contemporary art and architecture*. New Haven, CT: Yale University Press.

Yú, Q. (于青). (1987). 女人为奴的终结 [The end of females as slaves: On the legend of Zhāng Ailíng] In Yú Qīng & Jīn Hóngdá (Eds.), ≪张爱玲研究资料≫[Materials for studying Zhāng Ailíng] 海峡文艺出版社.

Yú, Q. (于青). (1991). 张爱玲传略. [A brief biography of Zhāng Ailíng] In Yú Qīng & Jīn Hóngdá (Eds.), ≪张爱玲研究资料≫ [Materials for studying Zhāng Ailíng] 海峡文艺出版社.

Zelizer, B. (1992). *Covering the body: The Kennedy assassination, the media, and the shaping of collective memory*. Chicago: University of Chicago Press.

Zhāng, A. (张爱玲). [1944] (2001). 自己的文章. [One's own papers] In Zǐ, Tóng (子通) & Yī, Qīng (亦清) (Eds.), ≪张爱玲评说六十年≫[Sixty years commentary and evaluation of Zhāng Ailíng] 中国华侨出版社.

Zhāng, W. (1999). *Confucianism and modernization: Industrialization and democratization of the Confucian regions*. New York: St. Martin's Press.

Zhāng, X. (1995). *The politics of aestheticization: Zhōu Zuòrén and the crisis of Chinese New Culture (1927–37)*. Unpublished doctoral dissertation, Duke University.

Zhāng, X. (张学智) (Ed.). (2005). 儒学与当代文明:纪念孔子诞生2555周年国际学术研讨会论文集 (卷一) [Confucianism and modern civilization: An international commemorative conference on the 2555th birthday] 九州出版社.

Zhāng, A., & Kingsbury, K. (2007). *Love in a fallen city*. [Qing cheng zhi lian] New York: New York Review of Books.

Zhāng, Júxiāng., & Zhāng, Tiěróng. (张菊香,张铁荣) (1986). Eds. 周作人研究资料 [Complied materials for the study of Zhōu Zuòrén] .天津人民出版社.

Zhang, F. F., Michaels, D. C., Mathema, B., Kauchali, S., Chatterjee, A., Ferris, D. C., et al. (2004). Evolution of epidemiologic methods and concepts in selected textbooks of the 20th century. *Sozial- und Praventivmedizin, 49*(2), 97–104.

Zhōu, G. (2002). *Language, myth, identity: The Chinese vernacular movement in a comparative perspective.* Unpublished doctoral dissertation, University of California at Davis.

Zhōu, R. (周仁政). (2002). 京派文学与现代文化. [Literature of the Peking School and modern culture] 长沙:湖南师范大学出版社.

Zhū, W. (朱文华). (1988). ≪胡适评传≫ [A biography of Hú Shì with commentary] 重庆:重庆出版社.

Zimmer, H. D., Mecklinger, A., & Lindenberger, U. (Eds.). (2006). *Handbook of binding and memory: Perspectives from cognitive neuroscience.* New York: Oxford University Press.

Zipser, D. (1991). Recurrent network model of the neural mechanism of short-term active memory. *Neural Computation, 3,* 179–193.

Zola-Morgan, S., Cohen, N. J., & Squire, L. R. (1983). Recall of remote episodic memory in amnesia. *Neuropsychologia, 21*(5), 487–500.

Index

Printed in the United States
by Baker & Taylor Publisher Services